Developments in Plant Genetics

Developments in Plant Genetics

Cheryl Natividad

Delve
PUBLISHING
www.delvepublishing.com

Developments in Plant Genetics

Cheryl Natividad

Delve Publishing
224 Shoreacres Road
Burlington, ON L7L 2H2
Canada
www.delvepublishing.com
Email: orders@arclereducation.com

ISBN: 978-1-77407-635-4 (Hardcover)

ABOUT THE AUTHOR

Cheryl Agdaca- Natividad graduated with a BS Biology degree (magna cum laude) at the Saint Louis University, Baguio City in the Philippines. She took the licensure examination for secondary teachers in 2003 where she ranked no. 10 among the examinees. She earned her masters degree in Genetics at the premier state university, University of the Philippines Los Baños. She taught in the same university from 2006-2017 handling lecture and laboratory classes in general biology, cell biology, genetics, and molecular genetics. Apart from teaching, her work involved proposing and revising laboratory manuals and mentoring students in their research work.

TABLE OF CONTENTS

LIST OF FIGURES

reddish-brown internal vascular discoloration. Transgenic lines expressing RGA2 (c, d) or *Ced9* (e, f) proved to be resistant to the fungus

Figure 3.21. First generation of transgenic lines (T_1) of jute (JBC1, JBC2, JBC3) with the *bar* gene were grown and then sprayed with the glufosinate herbicide Basta®. (Upper row) Non-transgenic, wild-type (WT) plants dies 6 days after the herbicide application while the transgenic plants recovered from the herbicide stress and grew new leaves and grew longer. (Lower row) Plants of T_2 generation also showed tolerance to Basta® application as opposed to the WT plants

Figure 3.22. Several steps are involved in the expression of stress tolerance, beginning from the perception of the drought stress and transducing the signal to transcription factors (TFs) for the expression of the genes involved in adaptation. Thus, drought tolerance is a complex phenotype and it is polygenic

Figure 3.23. (A) Transgenic rice plants (4–3 and 2–4) overexpressing *OsEREBP1* showed delayed onset of leaf rolling and wilting when subjected to drought stress compared to the non-transgenic plants (Kit). (B) Drought was imposed at the reproductive stage by withdrawing water during the panicle initiation stage until the plants were completely brown before they were rewatered. The transgenic rice plants showed new tillers which formed into panicles and grains indicating that they recovered from drought

Figure 3.24. (A) After being subjected to submergence stress, transgenic rice plants (4-3, 2-4), which overexpress *OsEREBP1*, showed flat and green leaves as opposed to the non-transgenic lines (Kit) which had rolled and yellowed leaves. (B) When compared to the transgenic lines, the non-transgenic, Kitaake lines showed higher length difference after submergence. (C) Transgenic lines were able to recover after prolonged submergence as indicated by the growth of new leaves, which were not observed in the non-transgenic lines. (D) After submergence, the leaves were stained with 3-3'diaminobenzidine tetrahydrochloride (DAB) and nitro-blue tetrazolium (NBT) to visualize $H_2O_2$and O_2^- accumulation, respectively. The intensity of brown and blue colors was greater in the non-transgenic lines compared to the transgenic ones

Figure 3.25. Transgenic lines of *Arabidopsis thaliana* (L12, L14), which overexpress *SOS1* gene from cotton, survived salt stress treatment (250 mM NaCl)for 7 days as opposed to the wild type plants (WT)

Figure 3.26. Three-week old transgenic (OE3, OE7, OE10) and wild type (WT) *Arabidopsis thaliana* plants were subjected to salt and drought stress conditions for 14 days. Results showed that significantly more transgenic plants survived than the WT

Figure 3.27. (A) Rice is a C3 plant where more than 90% of the total chloroplasts are found in the mesophyll cells (MCs) and very few chloroplasts are in the bundle sheath cells (BSCs). (B) On the other hand, *Setaria viridis* is a C4 plant so it exhibits the Kranz or wreath anatomy where MCs surround the BSCs and equal quantity of chloroplasts are in both cell types

Figure 3.28. In C3 photosynthesis, atmospheric CO_2 is assimilated by Ribulose-1,5-bisphosphate carboxylase oxygenase (Rubisco) in the mesophyll cell to form a 3-carbon intermediate. This whole process is called the Calvin cycle. Rubisco can also fix O_2

when there is more of this gas than CO_2 in the intercellular spaces of the leaf tissue. This results to photorespiration which wastes energy produced by photosynthesis. In C4 photosynthesis, the mesophyll cell contains phosphoenolpyruvate carboxylase (PEPC) which fixes atmospheric CO_2 generating a 4-carbon intermediate. This is decarboxylated to release CO_2 in the bundle sheath cell which contains Rubisco re-fixing it via the Calvin cycle. PEPC, unlike Rubisco, is insensitive to O_2 and so it concentrates CO_2around Rubisco, preventing photorespiration (Rizal et al., 2012)

Figure 3.29. Tobacco was genetically transformed to perform either one of three alternative photorespiratory pathways that metabolize glycolate in the chloroplast. Pathway 1 converts glycolate to glycerate using five genes from the *Escherichia coli* glycolate pathway that encode glycolate dehydrogenase, glyoxylate carboligase, and tartronic semialdehyde reductase. Pathway 2 and 3 converts glycolate to malate. Pathway 2 involved three transgenes encoding glycolate oxidase (from *Arabidopsis thaliana*), malate synthase (from pumpkin), and catalase (from *E. coli*). Catalase breaks down the hydrogen peroxide generated by glycolate oxidase. Finally, pathway 3 also contains pumpkin gene for malate synthase and *Chlamydomonas reinhardtii* glycolate dehydrogenase to avoid hydrogen peroxide generation when glycolate is converted to glyoxylate

Figure 3.30. During light stress, the xanthophyll cycle is induced which has a photoprotective role. This involves the removal of epoxy groups, as catalyzed by violaxanthin de-epoxidase (VDE), from violaxanthin producing the intermediate antheraxanthin, which is further de-epoxidized to form zeaxanthin. The reverse reaction which re-generates violaxanthin is catalyzed by zeaxanthin epoxidase (ZEP)

Figure 3.31. The nitrogen cycle converts nitrogen gas (N_2) into a form that can be assimilated by plants. This involves a nitrogen-fixing reaction performed by symbiotic bacteria in the root nodules of legumes. The resulting ammonium is converted to nitrites then nitrates by nitrifying bacteria. The nitrates are then taken up by the plants to be used as nitrogen source for certain biosynthetic processes. The nitrates can also be converted into N_2 by denitrifying bacteria. When the plant or certain plant parts die and decompose, the nitrogen-containing compounds in the plants are broken down into ammonia

Figure 3.32. The Golden Rice is genetically transformed in order to produce a significantly higher quantity of β-carotene in its endosperm. As a result, it has a golden yellow color as opposed to traditional rice

Figure 3.33. The synthesis of β-carotene in the Golden Rice endosperm involves three enzymes. The first one is phytoene synthase which synthesizes phytoene from geranylgeranyl diphosphate. Phytoene desaturase (crt1) catalyzes the next two reactions which produce lycopene. Lycopene cyclase catalyzes the final step of converting lycopene to β-carotene

Figure 3.34. (A) Folate or vitamin B_9 is a derivative of tetrahydrofolate, which is made up of pteridine, *para*-amino benzoic acid (pABA), and glutamate. (B) Pteridine is synthesized in the cytosol from GTP while pABA is made in the chloroplast (plastid).

These two precursors are then transported into the mitochondrion where they are used to synthesize tetrahydrofolate (THF). The glutamate tail is then added to THF

Figure 3.35. Canola is an edible variety of rapeseed that was developed by Canadian scientists. Its name is derived from Canada and "ola" (which means oil). Canola seeds (inset) are a major source of plant-based oil

Figure 3.36. Safflower is a oilseed crop with the most common varieties having yellow or orange flowers. Oil derived from its seeds (inset) contains a higher proportion of unsaturated fatty acids relative to saturated fatty acids

Figure 3.37. Overexpression of *SIMYB75* in tomato significantly increased anthocyanin levels relative to the wild type. This resulted to purple fruits at different growth stages and smaller, purple seeds

Figure 3.38. Production of viral glycoproteins as vaccine immunogens involves introducing the glycoprotein gene of the virus into an expression vector. The target plant is then genetically transformed (*Agrobacterium*-mediated in this case) with this recombinant vector. The glycoproteins produced in the transgenic plant are extracted, purified from contaminating host proteins, and then used for immunization to elicit neutralizing antibodies against the virus

Figure 4.1. Sequence-specific nucleases can induce double-stranded breaks (DSBs) within (a) protein coding genes, (b) non-protein coding genes or (c) regulatory elements. (d) The introduced DSBs may also result in the deletion of large chromosomal segments

Figure 4.2. A double-stranded break (DSB) of the target DNA that is induced by nucleases can be repaired by either non-homologous end-joining (NHEJ) or homology-directed repair (HDR). (A) The broken ends are joined together inaccurately in NHEJ-mediated repair, which is an error prone process resulting to either insertion or deletion of random nucleotides. This may lead to gene knockout or loss of function of the gene. HDR involves a donor DNA template with regions that are homologous to the sequences flanking the DSB. The DSB is repaired via homologous recombination which can introduce (B) precise gene modification or (C) gene insertion

Figure 4.3. A zinc-finger nuclease (ZFN) is designed by joining DNA-binding zinc-fingers to the DNA-cleavage domain of *Fok*1 endonuclease. Each zinc finger, represented by a circle, recognizes three nucleotides and so assembling three or four zinc finger motifs produces a sequence-specific DNA-binding domain. ZFNs function in pairs because *Fok*1 nuclease activity requires dimerization. The zinc finger-binding domain brings two *Fok*1 units together in the right orientation over the target sequence. This induces *Fok*1 dimerization and cleavage of the target sequence ensues

Figure 4.4. A pair of zinc-finger nucleases (ZFNs) recognize and bind specific DNA sequences via their zinc finger (ZF) domains. The dimerization of the *Fok*1 nuclease domains of the ZFN pair makes them cleave the DNA to create a double-strand break (DSB), which in turn initiates the DNA repair processes. The inaccurate non-homologous end-joining (NHEJ) can result in targeted mutagenesis. The homology-directed repair (HDR) occurs in the presence of a donor DNA with sequences homologous to the regions flanking the DSB. The donor DNA may be designed to carry a specific sequence

DNA

Figure 4.12. (A,E) The *Gn1a* knockout mutants (*gn1a—2* with a 2-bp deletion and *gn1a—10* with a 10 bp deletion) showed longer panicles than the wild type. (B,C) Their plant height is also significantly higher, and (D) they have more average numbers of flowers per panicle than the wild type. The other mutant line with a 3-bp deletion (*gn1a—3*), which resulted to a single amino acid deletion, showed similar phenotype in terms of panicle size, plant height, and flower number relative to the wild type

Figure 4.13. *DEP1* knockout mutants (*dep1–4* with 4-bp deletion and *dep1–2* with 2-bp deletion) had significantly shorter panicles (A,E) and plant height (B, C), but showed more flowers per panicle (D) compared to the wild type. Another mutant line, *dep1–3* with 3-bp deletion, served as a control which exhibited the same phenotype as the wild type

Figure 4.14. (A,B) The *GS3* knockout mutants with 4-bp deletion, *gs3-4*, and with 5-bp deletion, *gs3-5*, showed significantly longer grains compared to the wild type. This also resulted to higher grain weight in these mutants (C). The mutant with 9-bp deletion, *gs3-9*, had three amino acids deleted and served as the control with phenotypes like the wild type

Figure 4.15. The *IPA1* knock out mutants with 12-bp and 21-bp deletion, *ipa1–12* and *ipa1–21*, respectively, showed longer panicles (A,G), increased plant height (B, E) and more flowers (H) but lesser tiller number (F) compared with the wild type. Another mutant, *ipa1–5*, with a 5-bp deletion exhibited significantly reduced panicle length (C, G), a dwarf phenotype (D, E), lesser flowers (H) but greatly increased number of tillers (F) compared to the wild type. The mutant with 3-bp deletion, *ipa1–3*, showed wild type phenotypes (A, E-H)

Figure 4.16. Schematic of the structure of the transiently expressed CRISPR/Cas9 vector containing the sgRNA expression cassette and the Cas9 coding sequence. Five primer sets were designed to detect if the CRISPR/Cas9 DNA was integrated into the genome of the target plants

Figure 4.17. Overview of the genome-editing methods using transiently expressed CRISPR/Cas9 DNA or RNA. The designed constructs were introduced into immature embryos via particle bombardment and then the calli were regenerated in medium without herbicide selection. The presence of targeted mutation is determined via PCR-restriction enzyme digestion (PCR/RE) assay

Figure 4.18. There are several types of chimeric single guide RNAs (cgRNA) based on modality of function and orientation. (A) The *cis* repair template has the guide RNA (gRNA) (blue) attached directly to the repair template (green). The repair template is in antisense orientation if it is complementary to the non-target strand. It is in the sense orientation if, just like the gRNA, it is complementary to the target strand. (B) The *trans* repair template, has a tRNA (red) separating the gRNA from the repair template. It is cleaved at certain sites (indicated by arrows) to separate the two RNA molecules. Similar to the *cis* type, the *trans* repair template can either have an antisense or sense orientation relative to the gRNA. (Light blue line represents the scaffold RNA, green vertical lines represent the PAM sequence, red vertical lines represent the introduced mutations)

Figure 4.19. (A) Mutant rice lines with the *cis*-antisense cgRNA template with human scaffold RNA showed the desired point mutations (green nucleotides) in the *acetolactate synthase 1* (*ALS1*) gene indicating a successful homologous recombination between the target sequence in the genome and the repair template. Some insertions were also observed (red nucleotide). (B) Seeds from these mutant lines (L1 and L2) germinated on culture with the herbicide bispyribac sodium (BS) showing tolerance to this herbicide as opposed to the wild type (Ctrl)

Figure 4.20. CRISPR/Cas9-mediated gene editing was used to insert GOS2 promoter (GOS2 PRO) into the 5'-untranslated region (5'-UTR) of the *ARGOS8* gene. The gRNA was designed to guide the Cas9 to make a cut at the CRISPR-RNA target site 1 (CTS1). Homology-directed repair (HDR) of the cleavage using the template DNA containing the GOS2 PRO flanked by homology arms (HA1 and HA2) resulted to insertion of the desired promoter. This method was also used to replace entirely the native promoter of ARGOS8 (ARGOS8 PRO) with the GOS2 PRO. The main difference with the previous method was the Cas9-mediated cleavage of target sites (CTS2 and CTS3) flanking ARGOS8 PRO in order to remove it from the genome. GOS2 PRO is expected to increase the expression level of the *ARGOS8* coding sequence (CDS)

Figure 4.21. Multiplex gene editing via the CRISPR/Cas9 system can be accomplished by designing a polycistronic unit containing the codes for several gRNAs, Cas9, and/or RNA cleavage sequences (RCS). The stronger RNA polymerase II-dependent (Pol II) promoter is usually used to direct Cas9 expression while gRNA expression is driven by either a Pol II or RNA polymerase III-dependent (Pol III) promoter. The gRNAs can either (Ia) have their individual promoters or have (Ib) a single promoter to direct the expression of all the gRNAs in the unit. (IIa) A single promoter can also be used to control the expression of Cas9 and the gRNAs. (IIb) Another version of this strategy is to engineer the code for gRNA and RCS into the introns. (IIc) The single transcriptional unit can be simplified further by removing the RCS. (PolyA: polyadenylation sequences; 5'UT-R: 5' untranslated region)

Figure 4.22. (a) Cas9 can be modified to become a nickase, inducing single-stranded breaks in the target site. Therefore, a pair of nickases is required to generate a double-stranded break (DSB). (b) Cas9 may also be inactivated or rendered catalytically dead (dCas9) and then fused with *Fok*1 nuclease, which generates the DSB when it dimerizes with another *Fok*1 nuclease at the target site

Figure 4.23. The CRISPR/Cas13a system is composed of a crRNA that recognizes and binds the target RNA, which is then degraded by Cas13a

LIST OF TABLES

LIST OF ABBREVIATIONS

5'-UTR	5'-untranslated region
ABA	abscisic acid
ABI4	ABA-insensitive 4
ADCS	aminodeoxychorismate synthase
ADH	alcohol dehydrogenase
ADH1	alcohol dehydrogenase 1
AGI	Arabidopsis Genome Initiative
AID	arginine-rich intracellular delivery
ALS1	acetolactate synthase 1
AMA1	apical membrane antigen-1
ANT1	*anthocyanin mutant 1*
ASIC	application-specific integrated circuit
ATP	adenosine triphosphate
attL1	attachment site L1
BAC	bacterial artificial chromosome
BS	bispyribac sodium
BSCs	bundle sheath cells
BUSCO	Benchmarking Universal Single-Copy Orthologs
BXW	banana *Xanthomonas* wilt
CA	carbonic anhydrase
CAO1	chlorophyll A oxygenase 1
CCS	circular consensus sequence
CHN50	endochitinase-50
CLR	continuous long read
CMV	cauliflower mosaic virus
CpTI	cowpea trypsin inhibitor
CRISPR	clustered regularly interspaced short palindromic repeats
CSP	circumsporozoite protein

CTB	cholera toxin-B subunit
CTB-Pins	cholera toxin B subunit-proinsulin
CTS	cleavage of target sites
CTS1	CRISPR-RNA target site 1
DBG	de Brujin graph
DCL	dicer-like
DEP1	dense and erect panicle 1
DiT2	dicarboxylate transporter
DMH-11	dhara mustard hybrid-11
DNA	deoxyribonucleic acid
DREB1	drought responsive element binding 1
DSBs	double-stranded breaks
EBE	effector-binding element
EC	electrical conductivity
EC_e	electrical conductivity of the soil saturation extract
EFSB	eggplant fruit and shoot borer
EIC	Ebola immune complex
EPSPS	5-enolpyruvylshikimate 3-phosphate synthetase
ERF	ethylene response factors
ESTs	expressed sequence tags
FAD3A	fatty acid desaturase 3A
FBP/SBPase	fructose-1,6/sedoheptulose-1,7-biphosphatase
FW	fresh weight
GDC	glycine decarboxylase
GFP	green fluorescence protein
GLA	gamma-linolenic acid
GLK	golden2-like
GMO	genetically-modified organisms
Gn1a	grain number 1a
GOS2 PRO	GOS2 promoter
GOX	glyphosate oxidoreductase
GS	glutamine synthetase
GS2	glutamine synthetase 2
GS3	grain size 3

GSTs	glutathione-S-transferases
GTPCHI	GTP cyclohydrase I
GUS	*β*-glucuronidase
HA	hemagglutinins
HAL	halotolerance protein
HDR	homology-directed repair
HEN1	HUA enhancer1
HEPN	higher eukaryotes and prokaryotes nucleotide-binding domains
HKT	high-affinity potassium transporter
IgA	immunoglobulin A
IL-4	interleukin-4
IPA1	ideal plant architecture 1
IPCR	inverse polymerase chain reaction
IPK1	inositol-1,3,4,5,6-pentakiphosphate kinase 1
IPT	isopentenyl transferase
LA	linoleic acid
LEA	late embryogenesis abundant
LTB	labile enterotoxin B subunit
MAS	marker-assisted breeding
MCs	mesophyll cells
MDA	malondialdehyde
MEP	mesophyll envelope protein
MLO	mildew-resistance locus
MSD	maize streak disease
MSP-1	merozoite surface protein-1
MSV	maize streak virus
NADP-MDH	NADP-dependent malate dehydrogenase
NADP-ME	NADP-dependent malic enzyme
NBT	nitro-blue tetrazolium
NCED	9-*cis*-epoxycarotenoid dioxygenase
NGG	3-nucleotide G-rich
NGS	next-generation sequencing
NHEJ	non-homologous end-joining
NLS	nuclear localization signal

NORs	nucleolus organizing regions
NPQ	nonphotochemical quenching
NVCP	Norwalk virus capsid protein
OLC	overlap-layout-consensus
OMT1	2-oxoglutarate/malate transporter
OsERBP1	*Oryza sativa* ethylene-responsive binding protein 1
PAM	protospacer adjacent motif
PAT	phosphinothricin acetyltransferase
PCK	phosphoenolpyruvate carboxykinase
PCR	polymerase chain reaction
PCR/RE	PCR-restriction enzyme
PDC	pyruvate decarboxylase
PDR	pathogen-derived resistance
PEG	polyethylene glycol
PEPC	phosphoenolpyruvate carboxylase
PPDK	pyruvate orthophosphate dikinase
PPO	protoporphyrinogen oxidase
PPT1	PEP/phosphate transporter 1
PSI	photosystem I
PSII	photosystem II
PTDs	Protein transduction domains
QTL	quantitative trait locus
RCS	RNA cleavage sequences
RDR	RNA-dependent RNA polymerase
RGA2	resistance gene analog 2
RNA pol II	RNA polymerase II
RNA pol III	RNA polymerase III
RNA	ribonucleic acid
RNAi	RNA interference
RNPs	ribonucleoproteins
ROS	reactive oxygen species
RuBP	ribulose-1,5-bisphosphate
SAAT	sonication-assisted *Agrobacterium*-mediated transformation
SCMT	silicon carbide-mediated transformation

SDA	stearidonic acid
SMRT	single-molecule, real-time
SOS	salt overly sensitive
SSBs	single-stranded DNA breaks
StSy	stilbene synthase
TAC	transformation-competent artificial chromosome
TALE	transcription activator-like effector
TALEN	transcription activator-like effector nuclease
TFs	transcription factors
THF	tetrahydrofolate
TLP	trait landing pad
TPT	triose-phosphate translocator
TT4	transparent testa 4
TuMV	turnip mosaic virus
TuMV-GFP	TuMV expressing green fluorescent protein
UPS1	ureide permeases
UTR	untranslated region
UV-Vis	ultraviolet-visible
VDE	violaxanthin de-epoxidase
VInv	*vacuolar invertase*
VLPs	virus-like particles
WT	wild-type
YAC	yeast artificial chromosome
ZEP	*zeaxanthin epoxidase*
ZF	zinc finger
ZFNs	zinc-finger nucleases
ZMW	zero-mode waveguide detector

PREFACE

Plants are essential biotic components of the biosphere. Specifically, they have been key in supporting the life of mankind. It is not surprising then, if man has been fascinated with the science of plants and harnessed his scientific know-how and skills to improve on plant characteristics he finds useful and desirable. This becomes even more significant in the face of increasing stresses, both in frequency and magnitude, faced by plants.

The book *Developments in Plant Genetics* showcases the body of knowledge and applications that have accumulated regarding the study of the plant's DNA in the past two decades. It starts with the first chapter, which revisits the foundation genetic principles that can help the reader delve into the succeeding chapters. Apart from presenting the universal genetic concepts, the opening chapter also focuses on the idiosyncrasies of the plant genome, such as the presence of a chloroplast genome. Chapter 2 looks into the plant genome: its general and unique features as well as the workflow of genome sequencing, piecing the sequences together, and then mining meaningful biological information from the data. To date, there have been numerous plant genomes that have been sequenced but this chapter considers the plant model organism, *Arabidopsis thaliana*, as a case study.

Chapters 3 and 4 present how genetic principles have been used to improve endogenous traits or introduce non-inherent ones into plants. Chapter 3 describes the various techniques of genetic engineering that are employed in plants. Apart from the example of *A. thaliana*, this chapter also considers how various crops have been genetically modified in order to make them more resilient to either natural or man-made stressors, increase their yield, enhance their nutritional profile, and enable them to function as bioreactors for producing recombinant proteins. Chapter 4 discusses a relatively newer tool for genome manipulation, i.e., genome editing. It describes the procedure and application of the commonly used genome editing tools.

This book endeavors to speak to the reader using clear, simple language. Figures accompany the discussion of concepts allowing the reader to visualize information that are otherwise complicated or abstract. There are also tables that summarize, compare, or contrast information that are previously presented in the text. Great care was given to citing the proper sources and listing all references so that the reader knows where to look for additional, more in-depth discussion of the information presented in the book. These features of the book make it a useful reference material to the genetics student, the plant scientist, or the researcher. It can also be an interesting reading material for anyone, even those with no extensive genetics background, who wishes to learn about the recent developments in plant genetics.

Chapter 1

Introduction to Plant Genetics

CONTENTS

(Source: Sam Fentress, Creative Commons License)

The ground-breaking principles of inheritance proposed by Johann Gregor Mendel were based on his observations on plants, specifically the garden peas. Many genetic discoveries that followed were also results of experiments conducted on plants, which is understandable as using plants as experimental subjects poses less ethical dilemma than using animals or humans and less challenges logistically than using microorganisms. This chapter provides a recapitulation of the principles in genetics that will serve as springboard for the discussion in the successive chapters.

1.1. PLANT GENETIC MATERIAL

Plants, like all other organisms, have deoxyribonucleic acid (DNA) as their genetic material. They are eukaryotes which means that their DNA is enclosed by a nuclear membrane to form their nuclei. Their mitochondria and chloroplasts also have their own set of DNA. In some cases, segments of DNA may transfer from one organelle to the other (see Figure 1.1) (Chang et al., 2013).

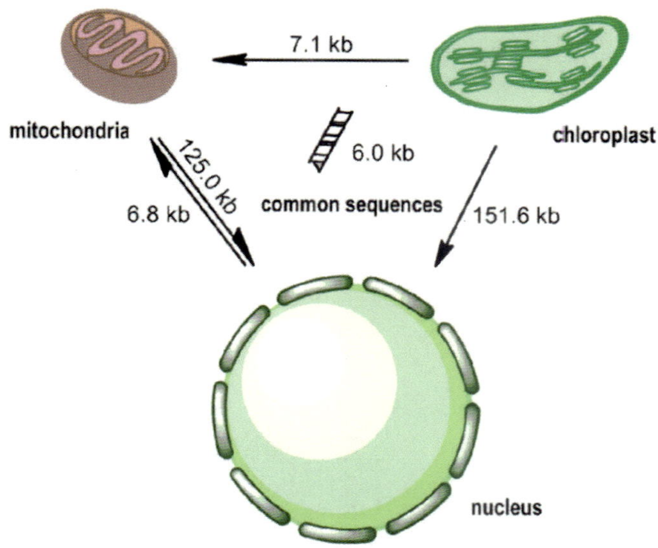

Figure 1.1: *DNA sequences may transfer among the nucleus, chloroplast, and mitochondrion, as in the case of soybean.*

Source: Chang et al. (2013).

1.1.1. Chemical Composition and Structure of Deoxyribonucleic Acid (DNA)

DNA was first isolated by Friedrich Mischer from human leukocytes. The substance he extracted was rich in phosphorus and resistant to proteolysis. Mischer initially called it nuclein in reference to the nuclei where the substance was located (Pray, 2008a). DNA is made up of repeating units of molecules called nucleotides. As a biomolecule, it is classified as a nucleic acid, along with ribonucleic acid (RNA). Nucleic acids are involved in storing, transmitting, and expressing biological information. Specifically, DNA is often described as the blueprint of an organism as it contains the code that determines the structure, function, as well as behavior of organisms.

A nucleotide is made up of three smaller units namely, a phosphate group, a pentose (i.e., a sugar with five carbon atoms), and a nitrogenous base (see Figure 1.2a). It was Phoebus Levene who deciphered the order of the components of a nucleotide (Pray, 2008a). The phosphate group is similar in both RNA and DNA. On the other hand, the sugar component of RNA is a ribose while in DNA, it is a 2-deoxyribose (see Figure 1.2c). The nitrogenous base can either be a purine, with two rings, or a pyrimidine, with a single ring in its chemical structure (see Figure 1.2b). There are two types of purine bases, adenine, and guanine, which can both occur in DNA and RNA molecules. On the other hand, there are three types of pyrimidine: cytosine that is observed in both types of nucleic acids, thymine that is unique to DNA, and uracil that is unique to RNA.

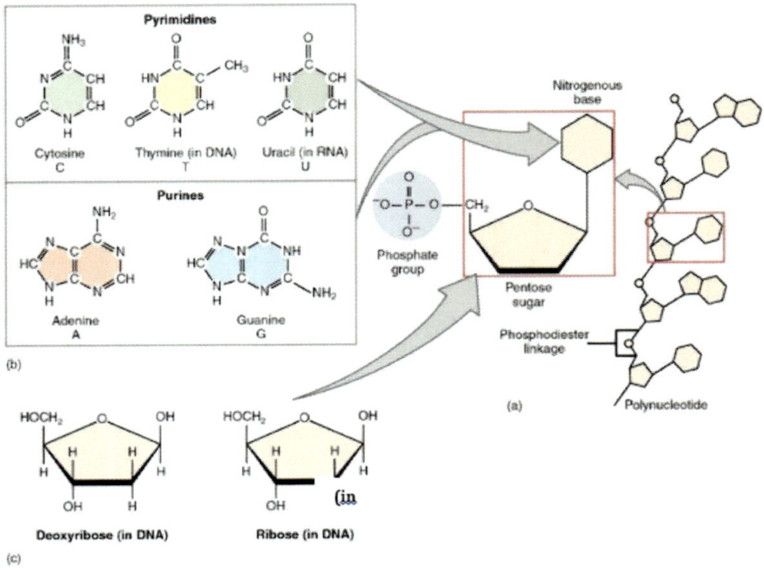

Figure 1.2: *(a) DNA and RNA are polynucleotides, i.e., made up of building blocks called nucleotides with phosphate group, pentose sugar, and nitrogenous base. (b) Nitrogenous bases may be classified as pyrimidines, which include cytosine, thymine, and uracil, or purines, which include adenine and guanine. (c) The five-carbon pentose sugar in DNA is deoxyribose while in RNA, it is ribose.*

Source: Rice University, Creative Commons License.

Several nucleotides are joined by a phosphodiester linkage (see Figure 1.2a) to form a polynucleotide. These nucleotides are similar in terms of their phosphate and sugar components and are distinct in their nitrogenous base. Therefore, a nucleotide is represented by the first letter of its nitrogenous base (e.g., A, G, C, T, U) so that the nucleotide sequence in the DNA is denoted by writing down the base sequence (e.g., CCCTTTAAGTCCGT).

The order of nucleotides along one strand is highly variable. Additionally, the strand displays polarity wherein one end with an exposed phosphate group is called the 3' end while the opposite end with the sugar is called the 5' end. In the case of RNA, it is made up of a single stranded polynucleotide. On the other hand, DNA is made up of two antiparallel strands, i.e., one is oriented 5' to 3' while the opposite strand is 3' to 5' (see Figure 1.3). The phosphate-deoxyribose backbone borders the DNA molecule while the nitrogenous bases are perpendicular to the backbone and the bases of the two strands that are facing each other are linked by hydrogen bonds. Thus, the entire length of the two strands of DNA is joined together by a series of hydrogen bonds (Figure 1.3).

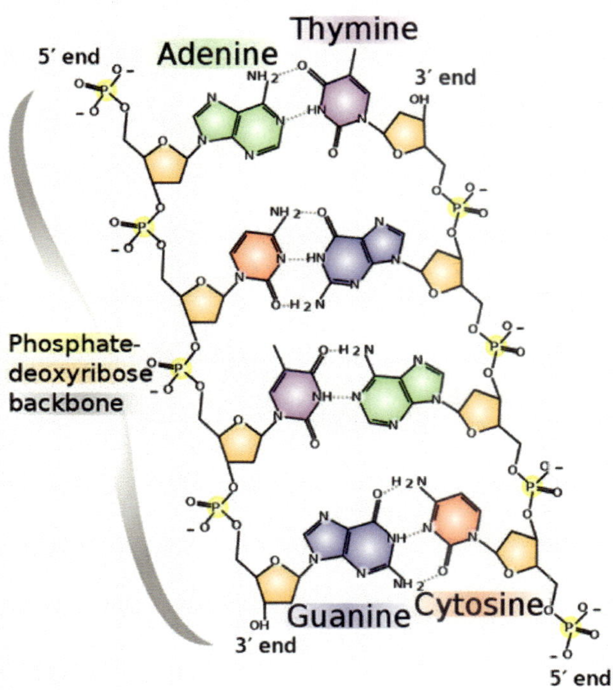

Figure 1.3: *DNA is a double-stranded molecule made up of antiparallel strands. The backbone of the molecule is made up of the repeating units of phosphate and deoxyribose while the nitrogenous bases are specifically paired and stacked in the interior region.*

Source: Madeleine Price Ball, Public Domain.

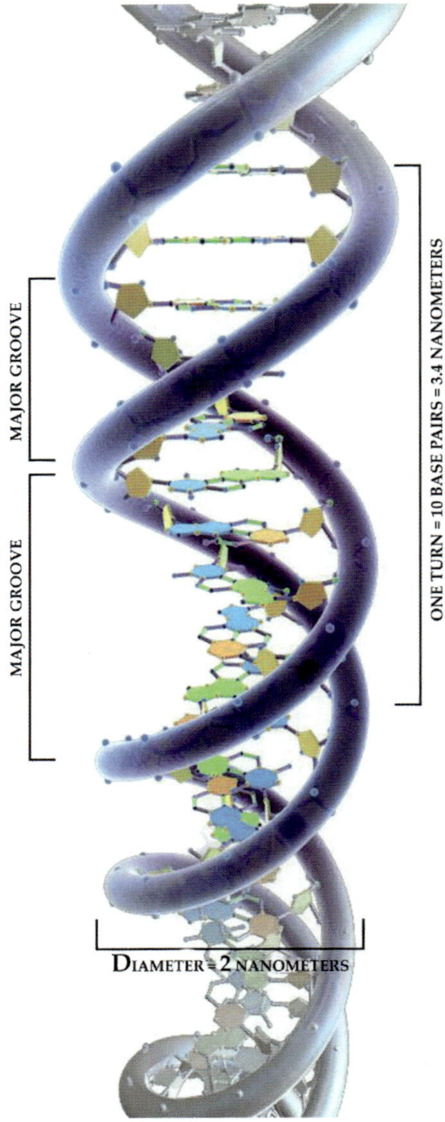

Figure 1.4: *The DNA model proposed by James Watson and Francis Crick is a double-stranded helix that rotates in a right-handed manner with a diameter of 2 nanometers. One complete turn has 10 pairs of nucleotides.*

Source: Zygote Media Group, www.3DScience.com, Creative Commons License.

1.1.2. DNA Is a Double Helix

The widely accepted model of the DNA is that it is a double-stranded helix with nucleotides (or to be more specific, bases) that show specific pairing. Erwin Chargaff discovered specific base pairing when he compared the DNA content of different species. He observed that regardless of the source organism or tissue, almost all DNA exhibit the same property, wherein the amount of adenine (A) is the same as the amount of thymine (T), and the amount of guanine (G) is similar to the amount of cytosine (C). This came to be known as Chargaff's rule (Pray, 2008a). This discovery formed the basis for the complementary base pairing exhibited in the DNA model proposed by James Watson and Francis Crick, where A is always paired with T and are linked via two hydrogen bonds while C pairs with G via three hydrogen bonds (see Figure 1.3).

The conclusion that DNA is a double-stranded helix is borne through the efforts of various scientists who performed x-ray diffraction studies on DNA samples. It was Maurice Wilkins who first broached the idea that DNA is helical based on his observations of the DNA with a reflecting microscope and its x-ray diffraction pattern. Raymond Gosling took a very clear x-ray diffraction photo which was used by Rosalind Franklin to conclude that DNA is a two-chain helix. The same photo as well as certain details of Franklin's work was the bases of Watson and Crick for them to propose that DNA is a double helix with the helices running in opposite directions (U.S. National Library of Medicine, 2017). Watson and Crick proposed that the DNA strands form right-handed helices (see Figure 1.4). They surmised that every 360° turn of the helix encompasses 10 pairs of nucleotides. The distance of the stacked nucleotides from each other is 3.4 Angstrom (Å) or 0.34 nanometer (nm). The outermost phosphate is located 10Å from the middle axis of the helix which means that the diameter of the helix is 20 Å or 2 nm (Watson and Crick, 1953).

1.1.3. Organization of the DNA

The DNA strands inside a cell can stretch up to a few meters long. In a eukaryotic cell like that of a plant, these very long structures are contained inside the microscopic nucleus, with a diameter that ranges from 1–20 micrometers (μm) (Smith, 1977; Annunziato, 2008). This indicates that the DNA is tightly packed and organized in order to fit inside the nucleus (see Figure 1.5). In the case of the plant cell, certain DNA molecules are located inside the mitochondrion and the chloroplast, which are organized quite differently from their nuclear counterpart.

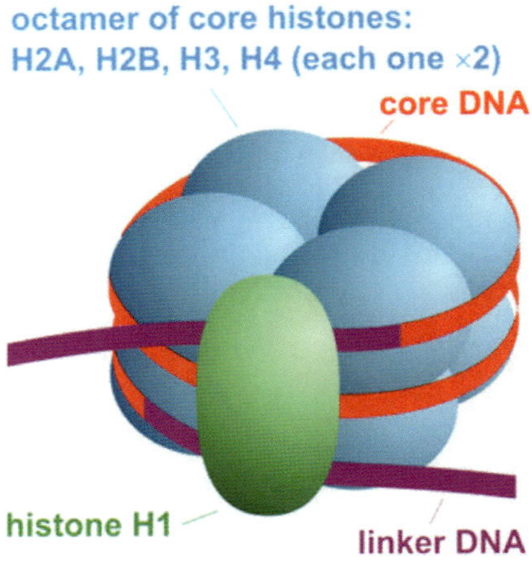

Figure 1.5: *DNA in eukaryotes is packaged into a more compact structure by five types of histone proteins. Two copies each of H2A, H2B, H3, and H4 form an octamer around which the DNA wraps around nearly twice. The resulting complex is called a nucleosome. H1 is bound to the linker DNA that joins adjacent nucleosomes together.*

Source: Darekk2, Creative Commons License.

1.1.4. Forming the Nuclear Chromosomes

Nuclear DNA is compacted in a eukaryotic cell by a family of proteins called histones. These are small, positively charged proteins that bind tightly to the DNA molecule via their electrostatic attraction to the negatively charged phosphate groups. There are five types of histones involved in DNA packaging, namely, H1, H2A, H2B, H3, and H4. Two copies each of H2A, H2B, H3, and H4 form an octamer, around which the DNA molecule wraps approximately 1.7 times (see Figure 1.5). Such DNA segment contains about 146 pairs of nucleotides or what are simply called, base pairs (bp). This DNA-protein complex makes up the nucleosome, which occurs repeatedly to form the chromatin. Therefore, the nucleosome is the structural and functional unit of the chromatin. The H1 protein wraps another 20 bp, which form part of the linker DNA. Considering that the chromosome is made up of millions of base pairs, each chromosome then has hundreds of thousands of nucleosomes joined by the so-called linker DNA, which are segments of

DNA not associated with the octamer. This gives the chromosome the beads-on-a-string appearance. The winding of DNA around the histones reduces its length by seven times (Annunziato, 2008), but thickens the diameter from 2 nm (which is the diameter of naked DNA) into 10 nm (see Figure 1.6).

The beads-on-a-string form of the chromatin folds further to form the 30-nm thick secondary structure (see Figure 1.6). This structure may influence the degree by which DNA-binding proteins access the DNA sequences (Iyer et al., 2011). This secondary structure packages the DNA further into a much shorter but thicker structure.

The 30-nm fiber forms loops to condense the chromatin further and form a linear chromosome (Iyer et al., 2011) (see Figure 1.6). This conformation occurs in the presence of the H1 histone. The chromosomes are replicated just before cell division and become highly condensed at metaphase stage of cell division, at which stage it has a diameter of about 1,400 nm (see Figure 1.6). At this point, the DNA has been shortened about 10,000 times (Annunziato, 2008).

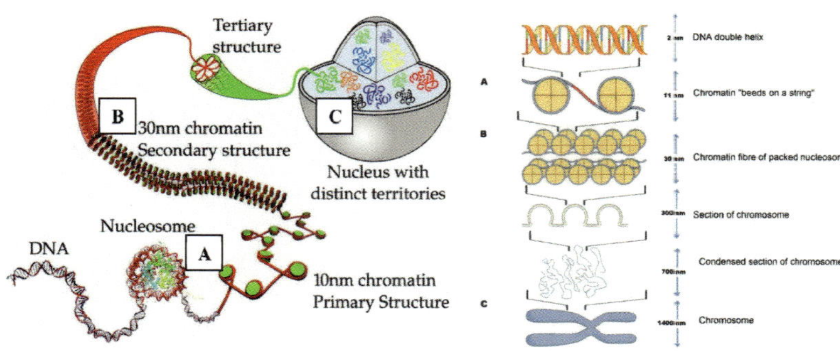

Figure 1.6: *The 2-nm DNA undergoes several levels of organization encompassing the (A) 10-nm nucleosome primary structure, the (B) 30-nm fiber secondary structure, which is further condensed into tertiary structure until the (C) chromosome is formed.*

Source: Left: Iyer, Kenward, and Arya (2011); Right: McAnena, Brown, and Kerin (2017).

1.1.5. The Organellar DNA

Most of the DNA is stored in the nucleus of eukaryotic cells but a small amount is found in the mitochondria, which are present in fungi, plants, animals, and humans, as well as in the chloroplasts of plants (Alberts et al., 2015; Lodish et al., 2003). These organellar DNA are significantly smaller in size compared to their nuclear counterpart and are circular in form rather than linear.

1.1.5.1. Mitochondrial DNA

As previously mentioned, the mitochondrion is an organelle found not just in plants but also in humans, animals, and fungi. It functions primarily to break down food molecules in order to generate energy, mostly in the form of adenosine triphosphate (ATP). Mitochondrial DNA (mtDNA) is located in the matrix of the mitochondrion. It is generally double-stranded and forms a closed circle (see Figure 1.7). Unlike the nuclear DNA, it is not bound by histones and thus, mtDNA is not highly condensed (Klug et al., 2005; Griffiths et al., 2000). In most organisms, mtDNA replicates at interphase so that a mitochondrion contains several mtDNA. The total amount of mtDNA varies among different cell types since it depends on factors that are variable among different cells, such as the number of mitochondria per cell, the number of mtDNA per mitochondrion, and the size of mtDNA (Lodish et al., 2003). This is in contrast to nuclear chromosomes that exist only as a single copy per cell, a condition known as haploid, or as two copies (diploid) (Griffiths et al., 2000).

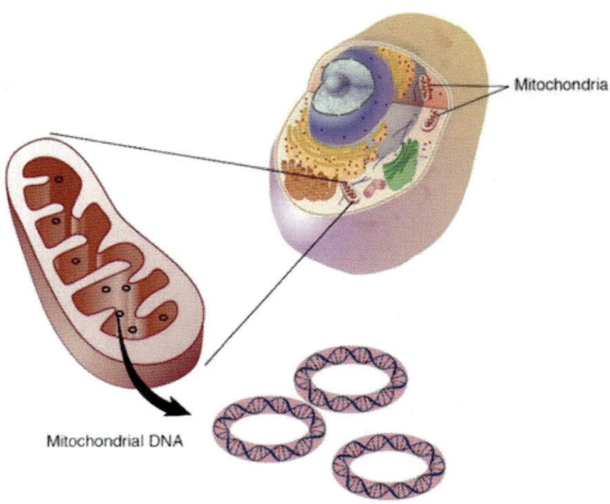

Figure 1.7: *Mitochondrial DNA is a double-stranded, circular structure that is located in the matrix of the mitochondrion. It usually occurs in multiple copies in each mitochondrion.*

Source: National Human Genome Research Institute, Public Domain.

While other eukaryotes have one type of mtDNA, plants have several types of mtDNA that cross-over with one another. Plant mtDNA is also significantly larger and more variable in size in contrast to those in other organisms (Lodish et al., 2003). For example, human, and animal mtDNA are made up of approximately 16,000 to 18,000 bp while mtDNA of yeast, which is a fungus, has 75 kilobase pairs (kbp). In contrast, the mustard plant has a mtDNA that is much bigger at 367 kbp (Klug et al., 2005).

1.1.5.2. Chloroplast DNA

The chloroplast is an organelle unique to organisms that perform photosynthesis such as algae, certain protists, and plants. It has machinery similar to mitochondrion that allows it to synthesize ATP molecules. However, unlike the mitochondrial ATPs that are utilized to drive various life-sustaining metabolic activities, chloroplast ATPs are used by the chloroplast itself to manufacture sugar.

Chloroplast DNA (cpDNA) is located in the stroma of the chloroplast, specifically in dark-staining areas called nucleoids (see Figure 1.8). It is double-stranded, circular, and not associated with any DNA packaging proteins just like mtDNA. It ranges in size from 100 to 225 kbp, making it larger than mtDNA. This is largely because cpDNA contains more noncoding regions and repetitive sequences than mtDNA (Lodish et al., 2003). Several copies of cpDNA may be present in each chloroplast and in some organisms; crossing-over among these may occur (Klug et al., 2005).

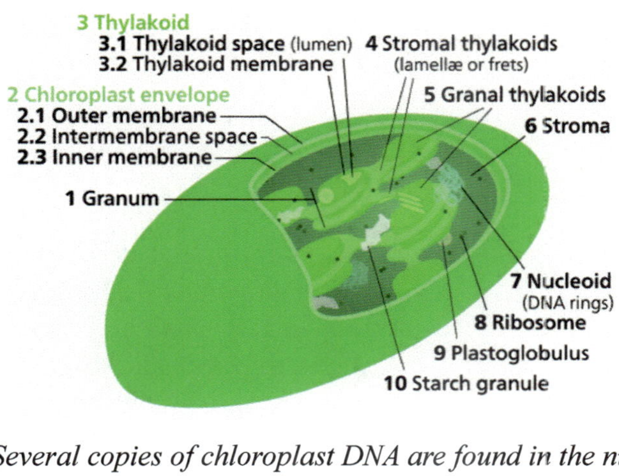

Figure 1.8: *Several copies of chloroplast DNA are found in the nucleoid in the stroma.*

Source: Kelvin Song, Creative Commons License.

1.2. EXPRESSING THE GENE

The total amount of DNA present per cell in an organism is referred to as its genome (U.S. National Library of Medicine, 2019a). Specifically, this refers to the haploid genome, i.e., the DNA content in one complete set of chromosomes. For example, most human cells have 46 chromosomes which are two sets (i.e., a diploid set) of 23 different chromosomes. Thus, the 3 billion bp comprising the human genome pertains to the DNA content of the 23 chromosomes (Pray, 2008b).

One chromosome may contain a very long DNA fiber equivalent to hundreds of thousands or millions of base pairs. During DNA replication, the entire genome sequence is faithfully copied, and the replicate is distributed to daughter cells during cell division. The heritability of the gene is due to the replicability of the DNA molecule. However, only certain segments of the DNA are considered genes. These are the DNA sequences that store information, which when expressed, produces RNA's, polypeptides, or proteins. These gene products, particularly the proteins, are the molecules that make up many of the structures in the cell. Moreover, enzymatic proteins catalyze the synthesis of non-protein compounds, the breakdown of substrates, and all other biochemical processes that are necessary to sustain life. It is in this sense that the genes are responsible for the structure, function, and behavior of the cell and the organism in general. These processes of DNA replication and gene expression are universal in the living world that these are denoted as the central dogma of molecular biology, which is represented by the following diagram:

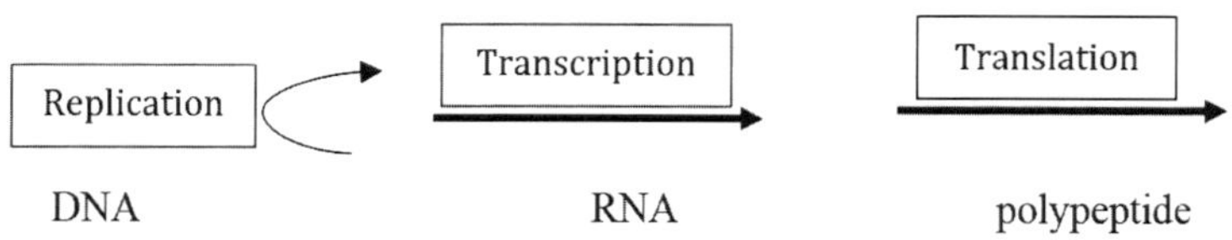

In the case of *Arabidopsis thaliana*, the first plant whose entire genome was sequenced, it has 125 million bp in its genome with an estimated 25,000 genes (Pray, 2008b). These constitute a very small percentage of the entire DNA content of the cell, which indicates that majority of the genome, is made up of non-coding DNA sequences. These were initially called "junk" DNA since they are not expressed into polypeptides or proteins. However, further researches revealed that these non-coding DNA play important roles in regulating gene expression (see Figure 1.9) or may form certain structures in the chromosome such as telomeres (U.S. National Library of Medicine, 2019b). The nuclear genes of eukaryotes and the chloroplast genes in plants are further subdivided into segments called exons and introns (see Figure 1.9). The exons are the actual coding sequences in a gene while the introns are the non-coding sequences between the exons (Alberts et al., 2015; Klug et al., 2005). In mammals, the genes in their mitochondrial genome have no introns and almost no non-coding, regulatory sequences (Lodish et al., 2003; Alberts et al., 2015).

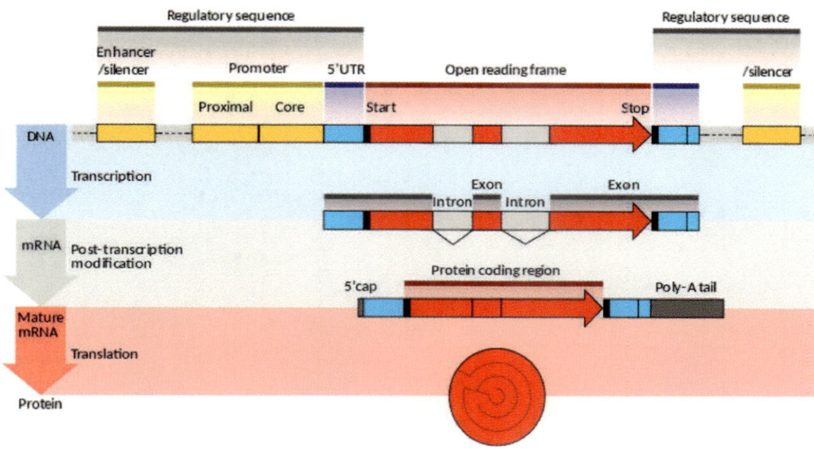

Figure 1.9: *The nuclear genes of eukaryotes have non-coding regions that do not specify any protein product but are necessary for controlling gene expression (e.g., enhancer/silencer, promoter, untranslated region (UTR)). The open reading frame has the coding sequence, specifically the exons, and intervening, non-coding sequences called the introns. The introns are removed from the RNA before the translation process.*

Source: Shafee and Lowe (2017).

1.2.1. Transcription Produces RNA from DNA

The complementary sequence of the gene, called the antisense strand, serves as the template during the process called transcription. The appropriate genetic or sense sequence is copied from the DNA into an RNA molecule (Alberts et al., 2015). This process involves certain proteins called transcription factors (TFs) and an enzyme called RNA polymerase. This enzyme can break the hydrogen bonds holding the two DNA strands together causing the helix to open up and unwind. It then lays down RNA nucleotides as it moves along the antisense sequence in a 5' to 3' direction (see Figure 1.10). It joins these RNA nucleotides by catalyzing the formation of phosphodiester linkages. RNA polymerase follows specific base pairing as it synthesizes RNA, i.e., it pairs G in the template with C (and vice-versa), T in the template with A, and A in the template with U. Therefore, the genetic sequence (sense strand) and the RNA sequence are similar to each other except that the RNA molecule has U's in place of the T's in the gene (see Figure 1.10). When transcription is terminated, the RNA molecule separates from its DNA template, allowing the two DNA strands to be rejoined into a helix.

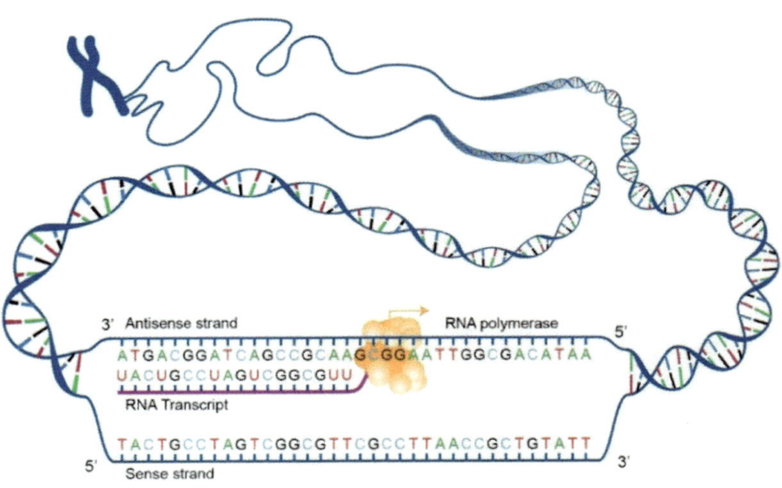

Figure 1.10: *In transcription, an RNA copy of the gene is made by the specific base pairing of the sequences in the antisense strand. The synthesis of the RNA transcript is catalyzed by RNA polymerase.*

Source: National Human Genome Research Institute, Public Domain.

There are three types of RNA molecules formed after transcription namely, messenger RNA (mRNA), transfer RNA (tRNA), and ribosomal RNA (rRNA). These are all needed in the next stage, the synthesis of polypeptide (see Figure 1.14). Every three nucleotides in the linear, single-stranded mRNA, which is read from 5' to 3' direction, are called a codon. It is equivalent to a certain amino acid (see Figure 1.11). In the case of eukaryotic mRNA, the introns which do not contain codons are removed prior to translation so that only the exons are left in the mRNA (see Figure 1.9). There are 64 codons, 61 of which specify the 20 amino acids. Among these 61, AUG is considered to be the start codon as it is the first codon to be translated during protein synthesis. The other three codons (UGA, UAA, UAG) do not correspond to any amino acid. Instead, these initiate the termination of the translation process and so these are called stop codons. Since there are more codons than amino acids, it follows that some amino acids may be specified by more than one type of codon. In this sense, the genetic code is said to be degenerate.

First letter	Second letter: U	Second letter: C	Second letter: A	Second letter: G	Third letter
U	UUU Phe	UCU Ser	UAU Tyr	UGU Cys	U
U	UUC Phe	UCC Ser	UAC Tyr	UGC Cys	C
U	UUA Leu	UCA Ser	**UAA Stop**	**UGA Stop**	A
U	UUG Leu	UCG Ser	**UAG Stop**	UGG Trp	G
C	CUU Leu	CCU Pro	CAU His	CGU Arg	U
C	CUC Leu	CCC Pro	CAC His	CGC Arg	C
C	CUA Leu	CCA Pro	CAA Gln	CGA Arg	A
C	CUG Leu	CCG Pro	CAG Gln	CGG Arg	G
A	AUU Ile	ACU Thr	AAU Asn	AGU Ser	U
A	AUC Ile	ACC Thr	AAC Asn	AGC Ser	C
A	AUA Ile	ACA Thr	AAA Lys	AGA Arg	A
A	AUG Met	ACG Thr	AAG Lys	AGG Arg	G
G	GUU Val	GCU Ala	GAU Asp	GGU Gly	U
G	GUC Val	GCC Ala	GAC Asp	GGC Gly	C
G	GUA Val	GCA Ala	GAA Glu	GGA Gly	A
G	GUG Val	GCG Ala	GAG Glu	GGG Gly	G

Figure 1.11: *The genetic code chart lists the 64 codons that may occur in a messenger RNA, 61 of these specify amino acids (designated via the three letter abbreviations of their names) with AUG being the start codon. The remaining three, UAA, UAG, and UGA, are non-coding stop codons.*

Source: OpenStax CNX from National Institutes of Health, Public Domain.

The tRNA exhibits a certain degree of intramolecular base pairing so that it assumes a folded, secondary cloverleaf structure with stems and loops (see Figure 1.12). It is responsible for bringing the amino acid to the site of protein synthesis. Its 3’ end is where the amino acid is attached. In one of its loops, three nucleotides are protruding which are called anticodon. These are complementary to a certain codon in the mRNA. Therefore, the amino acid carried by the tRNA corresponds to the codon, which complements that tRNA’s anticodon (see Figure 1.12). However, specific base pairing between the codon and the anticodon is not observed in the codon's third nucleotide. This is called the wobble position. This is why there is a smaller number of tRNAs that correspond with the 61 protein-coding codons in the mRNA. For example, a tRNA with G in the wobble position can base pair with either C or U so that the phenylalanine codons UUU and UUC are both recognized by a tRNA with the anticodon GAA (Alberts et al., 2015).

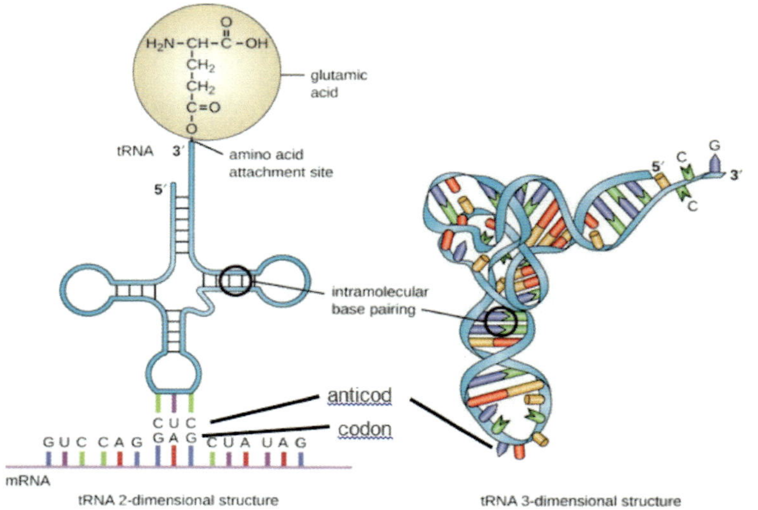

Figure 1.12: *The transfer RNA (tRNA) is single-stranded but with certain regions that can exhibit base pairing with each other resulting to its cloverleaf configuration. Its 3'end is the site for attaching amino acid, which in this case is glutamic acid because it is specified by the GAG codon in the messenger RNA (mRNA) that pairs with the CUC anticodon of the tRNA.*

Source: OpenStax CNX, Creative Commons License.

The rRNA is called such since it associates with ribosomal proteins to form the ribosomes, which are ribonucleoprotein complexes where protein synthesis occurs. These have large and small subunits (see Figure 1.13). Specifically, the ribosomes of eukaryotes have the 60 Svedberg units (S) subunit and the 40S small subunit. The whole structure come together to form the 80S ribosome, which is larger and denser than the 70S prokaryotic ribosome.

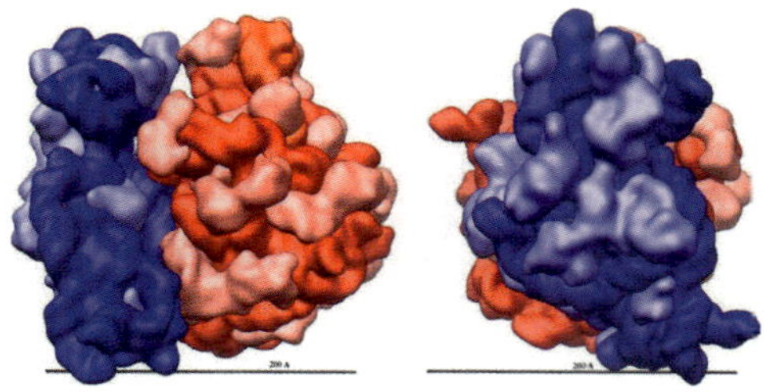

Figure 1.13: *The ribosome has a large subunit (red) and a small subunit (blue). Each subunit is a complex made up of ribosomal proteins (light blue and pink) and ribosomal RNA (rRNA) (dark blue and red).*

Source: Vossman, Creative Commons License.

1.2.2. Translation Produces Protein

As mentioned previously, the three types of RNA molecules produced after transcription are necessary for translation, the process of manufacturing protein from the instructions copied from the gene. This occurs in the cytoplasm so this means that these RNAs must migrate from the nucleus, where they are produced, to the cytoplasm. Ribosomes contain sites for the entry of incoming tRNA bearing amino acids as well as exit sites of empty tRNAs (see Figure 1.14). These also contain the enzyme peptidyl transferase that catalyzes the formation of peptide bonds that link together the amino acids to form a polypeptide.

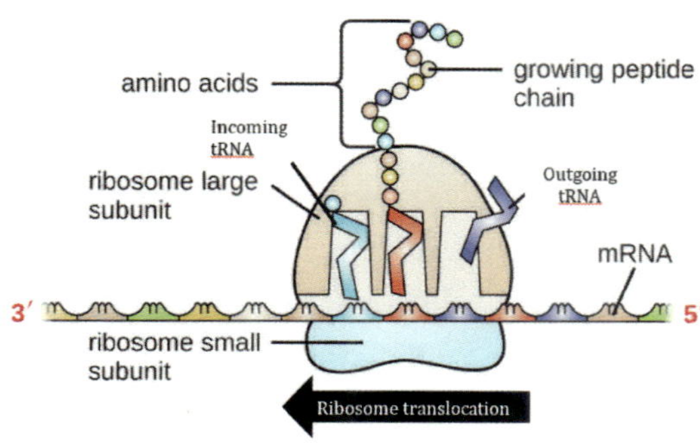

Figure 1.14: *The messenger RNA (mRNA) is the template that contains the codons for the amino acid sequence of the polypeptide to be synthesized. The transfer RNA (tRNA) carries the amino acid that corresponds to each codon in the mRNA while the ribosome links the amino acids to form the polypeptide. The ribosome moves along from the 5' end of the mRNA going to the 3' direction.*

Source: OpenStax CNX, Creative Commons License.

As stated in the previous section, the first codon in the mRNA to be translated is AUG at the 5' end which specifies the amino acid, methionine. Then the ribosome moves to the 3' direction so that the next codon is at the entry site in the ribosome while the met-tRNA is at the exit site. The tRNA bearing the amino acid corresponding to the second codon enters the translation complex. The peptidyl transferase links the two amino acids with a peptide bond to form a dipeptide that is attached to the second tRNA. The first tRNA at the exit site, now devoid of its amino acid, leaves the translation complex (see Figure 1.14). These series of steps recur as the ribosome is translocated until all the codons in the mRNA have been translated, leading to a lengthening polypeptide. When one of the stop codons is encountered, the process is terminated. This entails the dissociation of the polypeptide chain from the translation complex and the separation of the two ribosome subunits from each other.

The immediate product of the translation process is a polypeptide. In order to become a functional protein, this must undergo further processing, such as the removal of certain amino acids in the chain or the addition of chemical compounds like an oligosaccharide, phosphate, or methyl. The polypeptide also folds to achieve a certain secondary or tertiary structure. In some cases, the polypeptide is simply a subunit of a much bigger protein molecule.

1.2.3. Protein Products of Organellar Genomes

The nuclear genome stores the code for most of the proteins in the cell. Many of these proteins are eventually exported from the cytoplasm into the various organelles such as the mitochondrion and the chloroplast. However, some of the proteins involved in ATP synthesis such as those in the electron transport chain and certain subunits of the ATP synthase complex are encoded in the mitochondrial genome (see Figure 1.15). Considering that the mitochondrion has its own set of ribosomes, these proteins are synthesized within the mitochondrion. The tRNAs, rRNAs, and proteins needed for mitochondrial translation are also encoded in the mtDNA (Griffiths et al., 2000; Lodish et al., 2003) (see Figure 1.15). Plants have genes in their mtDNA that code for a 5S mitochondrial rRNA. This is found only in the ribosomes of plant mitochondria (Lodish et al., 2003).

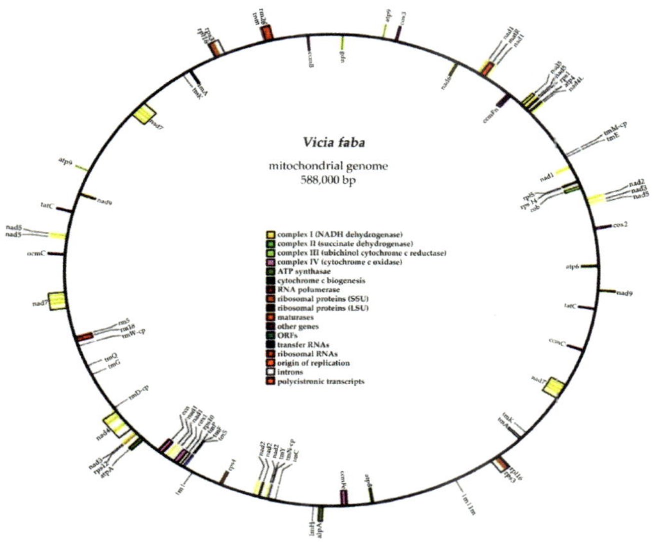

Figure 1.15: *The mitochondrial genome of the legume Vicia faba contains genes for certain proteins in the complexes for electron transport, ATP synthase subunits, RNA polymerase, ribosomal proteins as well as rRNAs and tRNAs.*

Source: Negruk (2013).

There are some differences in terms of the amino acids specified by certain codons in the nuclear genome and the mitochondrial genome of animals and fungi. Variation is also observed in the code of the mitochondrial genome from different species (Griffiths et al., 2000; Lodish et al., 2003). For instance, UGA is normally a stop codon in the nuclear genome. This is also recognized as a stop codon in the plant mitochondrial genome, but it is translated as tryptophan in the human and fungi mitochondrial genome (Lodish et al., 2003).

Plants have an additional genome in their chloroplast. This genome normally encodes 80–90 proteins and approximately 45 RNAs (Alberts et al., 2015). It stores the code for the RNA polymerase subunits, tRNAS, rRNAs, and ribosomal proteins that are involved in transcription of cpDNA and protein synthesis involving the chloroplast ribosomes (see Figure 1.16). Other genes in the genome code for some proteins involved in photosynthesis, such as two subunits of ribulose 1,5-bisphosphate carboxylase (Rubisco), as well as in ATP synthesis like components of the electron transport chain and ATP synthase complex subunits (Griffiths et al., 2000; Lodish et al., 2003) (see Figure 1.16). In the liverwort, a number of chloroplast genes specify proteins that are part of the thylakoid membrane (Klug et al., 2005). Synthesis of these proteins follows the same code as the nuclear genome (Lodish et al., 2003). However, the start codon, AUG, in the chloroplast genome is read as *N*-formylmethionine, just like in prokaryotes, instead of the usual methionine (Alberts et al., 2015).

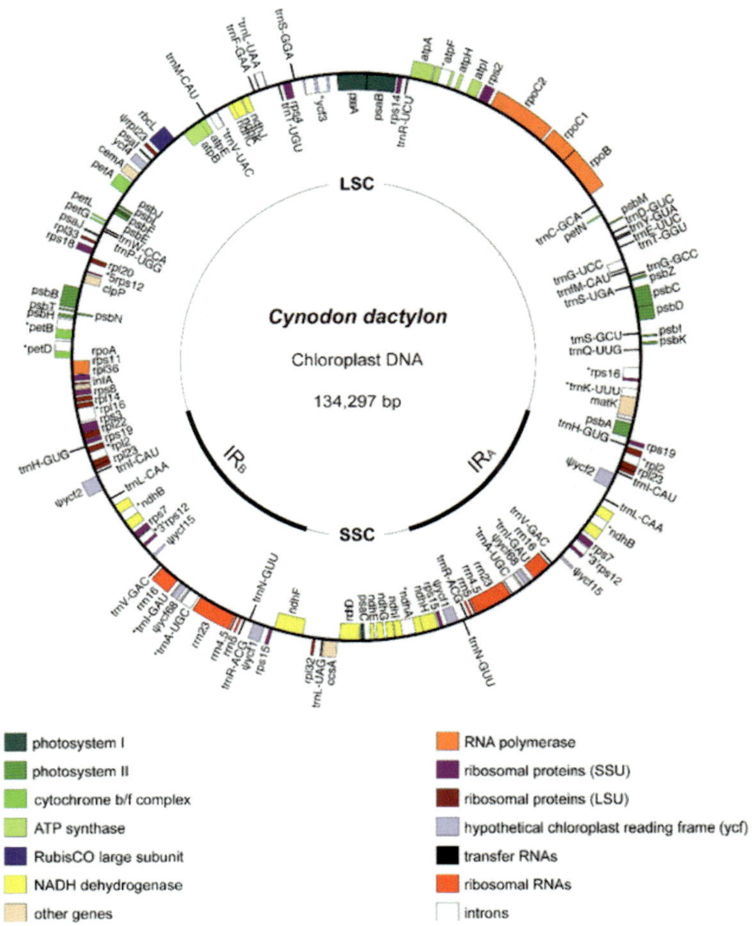

Figure 1.16: *The chloroplast genome of Bermuda grass, Cynodon dactylon, contains genes for tRNAs, rRNAs as well as proteins involved in transcription (e.g., RNA polymerase), translation (e.g., ribosomal proteins), ATP synthesis (e.g., photosystems, ATP synthase, NADH dehydrogenase), and photosynthesis (e.g., Rubisco large subunit).*

Source: Huang, Cho, Haryono, and Kuo (2017).

REFERENCES

1. Alberts, B., Johnson, A., Lewis, J., Morgan, D., Raff, M., Roberts, K., Walter, P., Wilson, J., & Hunt, T., (2015). Molecular Biology of the Cell (6th edn.). Garland Science: New York.
2. Annunziato, A. T., (2008). DNA packaging: Nucleosomes and chromatin. Nature Education, 1(1), 26.
3. Chang, S., Wang, Y., Lu, J., Gai, J., Li, J., Chu, P., Guan, R., & Zhao, T., (2013). Mitochondrial genome of soybean reveals complex genome structures and gene evolution at intercellular and phylogenetic levels. PLoS One, 8(2), e56502. https://doi.org/10.1371/journal.pone.0056502 (Accessed on 18 December 2019).
4. Darekk2. Creative Commons Attribution-Share Alike 3.0 Unported License. https://commons.wikimedia.org/wiki/File:Nucleosome_organization.png (Accessed on 18 December 2019).
5. Griffiths, A. J. F., Miller, J. H., Suzuki, D. T., Lewontin, R. C., & Gelbart, W. M., (2000). An Introduction to Genetic Analysis (7th edn.). W. H. Freeman: New York.
6. Huang, Y. Y., Cho, S. T., Haryono, M., & Kuo, C. H., (2017). Complete chloroplast genome sequence of common Bermuda grass (*Cynodon dactylon* (L.) Pers.) and comparative analysis within the family Poaceae, Creative commons attribution 4.0 international license. PLoS One, 12(9), e0184409. https://doi.org/10.1371/journal.pone.0179055 (Accessed on 18 December 2019).
7. Iyer, B. V. S., Kenward, M., & Arya, G., (2011). Hierarchies in eukaryotic genome organization: Insights from polymer theory and simulations. Creative commons attribution 2.0 generic license. BMC Biophysics 4, 8. https://doi.org/10.1186/2046-1682-4-8 (Accessed on 18 December 2019).
8. Kelvin Song, Creative Commons Attribution 3.0 Unported License. https://commons.wikimedia.org/wiki/File:Chloroplast_mini.svg (Accessed on 18 December 2019).
9. Klug, W. S., Cummings, M. R., & Spencer, C. A., (2005). Concepts of Genetics (8th edn.). Benjamin Cummings: San Francisco.
10. Lodish, H., Berk, A., Matsudaira, P., Kaiser, C. A., & Krieger, M., (2003). Molecular Cell Biology (5th edn.). W. H. Freeman and Co.: New York.

11. Madeleine Price Ball. Public Domain, https://commons.wikimedia.org/wiki/File:DNA_chemical_structure.svg (Accessed on 18 December 2019).
12. McAnena, P., Brown, J. A. L., & Kerin, M. J., (2017). Circulating nucleosomes and nucleosome modifications as biomarkers in cancer. Creative commons attribution 4.0 international license. Cancers, 9(1), 5. https://doi.org/10.3390/cancers9010005 (Accessed on 18 December 2019).
13. National Human Genome Research Institute. Public Domain. https://www.genome.gov/genetics-glossary/Transcription; https://www.genome.gov/genetics-glossary/Mitochondrial-DNA?id=129 (Accessed on 18 December 2019).
14. Negruk, V., (2013). Mitochondrial genome sequence of the legume *Vicia faba*. Creative commons attribution 3.0 unported license. Frontiers in Plant Science 4, 128. https://doi.org/10.3389/fpls.2013.00128 (Accessed on 18 December 2019).
15. OpenStax CNX from National Institutes of Health. Public Domain. https://history.nih.gov/exhibits/nirenberg/HS5_cracked.htm (Accessed on 18 December 2019).
16. OpenStax CNX. Creative Commons Attribution 4.0 International License. https://cnx.org/contents/5CvTdmJL@4.4:hRuNts1p@3/Structure-and-Function-of-RNA (Accessed on 18 December 2019).
17. Pray, L. A., (2008a). Discovery of DNA structure and function: Watson and Crick. Nature Education, 1(1), 100.
18. Pray, L. A., (2008b). Eukaryotic genome complexity. Nature Education, 1(1), 96.
19. Rice University, Creative Commons Attribution 4.0 International License. https://opentextbc.ca/anatomyandphysiology/chapter/2-5-organic-compounds-essential-to-human-functioning/ (Accessed on 18 December 2019).
20. Sam Fentress. Creative Commons Attribution-Share Alike 2.0 Generic License. https://en.wikipedia.org/wiki/File:Corncobs.jpg (Accessed on 18 December 2019).
21. Shafee, T., & Lowe, R., (2017). Eukaryotic and prokaryotic gene structure. Creative Commons Attribution 4.0 International License, Wiki Journal of Medicine, 4(1). doi: 10.15347/wjm/2107.002.

22. Smith, H., (1977). The Molecular Biology of Plant Cells (Vol. 14, p. 2013). University of California Press: Berkeley. http://ark.cdlib.org/ark:/13030/ft796nb4n2/ (Accessed on 18 December 2019).

23. U.S. National Library of Medicine, (2017). Profiles in Science: The Rosalind Franklin Papers. https://profiles.nlm.nih.gov/ps/retrieve/Narrative/KR/p-nid/187 (Accessed on 18 December 2019).

24. U.S. National Library of Medicine, (2019a). Genetics Home Reference: What is a Genome? https://ghr.nlm.nih.gov/primer/hgp/genome (Accessed on 18 December 2019).

25. U.S. National Library of Medicine, (2019b). Genetics Home Reference: What is Noncoding DNA? https://ghr.nlm.nih.gov/primer/basics/noncodingdna (Accessed on 18 December 2019).

26. Vossman, Creative Commons Attribution-Share Alike 3.0 Unported License. https://commons.wikimedia.org/wiki/File:Ribosome_shape.png (Accessed on 18 December 2019).

27. Watson, J. D., & Crick, F. H. C., (1953). Molecular structure of nucleic acids: A structure of deoxyribose nucleic acid. Nature, 171, 737–738.

28. Zygote Media Group, Creative Commons Attribution 2.5 Generic License. www.3DScience.com (Accessed on 18 December 2019).

Chapter 2

The Plant Genome

CONTENTS

(From: (Left) Marie-Lan Nguyen, Creative Commons License; (Right) Abizar, Creative Commons License).

A genome refers to a complete set of chromosomes and is designated as X. Thus, an organism with two chromosome sets has 2x and is called diploid, and those with more than two are called polyploids with 3X, 4X, and so on. On December 13, 2000, an international consortium of scientists successfully completed sequencing the first plant genome, that of *Arabidopsis thaliana* (see opening image). In 1996, the Arabidopsis Genome Initiative (AGI) was formed and within a span of four years, they were able to determine the entire sequence of the DNA in the plant's five chromosomes (National Science Foundation, 2000). This chapter describes the beginning of genome sequencing as well as the recent developments in terms of the technology and as the information generated from sequencing efforts in plants.

2.1. THE SIZE OF THE PLANT GENOME

The plant kingdom shows immense diversity, and this is reflected at the genetic level where genome size is highly variable even for members of the same family. These differences determine the feasibility of genomics effort, the methodology employed, the amount of data gathered, and the interpretation of such data.

2.1.1. DNA Content

The DNA content of the nucleus is commonly designated as the C value. Thus, the DNA content of a haploid cell, such as a gamete, is 1C while that of a diploid is 2C. Several plants are polyploid and so these may have a much higher C value. Different tissues of a single plant may have different DNA content. For example, the cotyledons of peas are polyploid with DNA content greater than 2C since their cells have undergone several rounds of endoreduplication (Cullis, 2004).

The genome that is conventionally sequenced is the 1C genome since it is the minimum amount of DNA to contain the complete set of genes (Peterson and Arick II, 2018). It is quantified either in terms of number of base pairs per haploid nucleus or the mass of DNA in picograms (pg). These two values can be interconverted by using the actual conversion of 1 pg = 980 megabase pairs (Mbp) (Cullis, 2004). Another way of converting DNA quantity to various units is by using the 1-2-3 method, i.e., 1 micrometer (μm) of linear double-stranded DNA approximately weighs 2 million Daltons and contains about 3 kbp of DNA (Hodgkin, 2001). Apart from being an indicator of genome size, the C value is related to the plant's life history, geographical distribution, growth response to certain environmental factors, and resistance to radiation (Cullis, 2004).

The most widely used method for estimating genome size is flow cytometry. This involves extracting nuclei and staining them with a fluorescent quantitative DNA stain (e.g., propidium iodide). The stained nuclei move into a thin stream of liquid one at a time. The liquid stream passes through a laser, which activates the fluorescent DNA stain in the nuclei. As a result, each nucleus gives off fluorescence that is proportional to its DNA content. A sample with a known genome size is used as a standard to be able to infer the genome size of the sample of interest (see Figure 2.1). This is the easiest method of estimating genome size as it can obtain measurements from thousands of nuclei in just a few minutes (Peterson and Arick II, 2018).

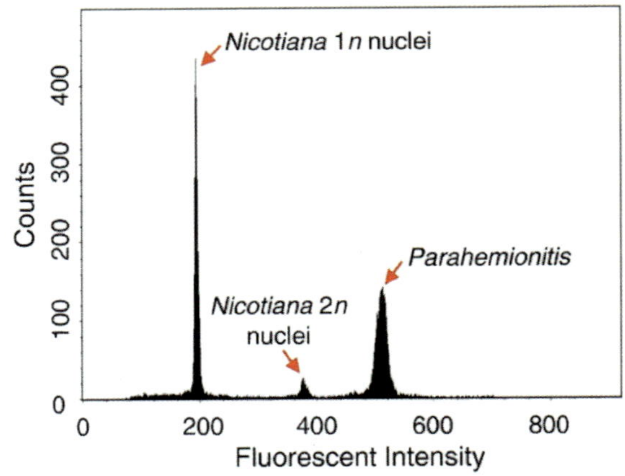

Figure 2.1: *An example of a flow cytometry result to determine the genome size of the fern Parahemionitis cordata with a haploid (1n) and diploid (2n) nuclei of Nicotiana serving as standards.*

Source: Li and Harkess (2018).

The nuclear genomes of plants range in size from 125 Mbp just like in *Arabidopsis thaliana*, crucifers, roses, certain trees, some pteridophytes, and bryophytes up to much larger genome size with more than 70 gigabase pairs (Gbp) in certain pteridophytes and angiosperms (Heslop-Harrison, 2003). Extremely small genomes have also been observed such as in *Genlisea aurea* (corkscrew plant) with a mere 60 Mbp while *Paris japonica* (a rare Japanese plant) exhibits a staggering 152 Gbp genome size (Michael, 2014). This variation in DNA content is also observed within genus. For example, members of the genus *Rosa* exhibit more than 11-fold variation in genome size. The genome size of some plants representing algae, monocots, and dicots are listed in Table 2.1.

Table 2.1: *Comparison of Some of the Sequenced Plant Genomes*

Plant Name	Chromosome Number	Genome Size	Number of Coding Genes
Actinidia chinensis var. *chinensis* (kiwi fruit)	2n = 58	554 Mbp	33,044
Chlamydomonas reinhardtii (unicellular green algae)	n = 17	120 Mbp	17,743
Glycine max (soybean)	2n = 40	1.115 Gbp	55,897

Helianthus annuus (sunflower)	2n = 34	3.6 Gbp	52,191
Musa acuminata (banana)	2n = 22/3n = 33	473 Mbp	36,525
Oryza sativa Indica (rice)	2n = 12	427 Mbp	40,745
Solanum tuberosum (potato)	2n = 24	840 Mbp	39,021
Triticum aestivum (bread wheat)	2n = 6x = 42	17 Gbp	107,891
Vitis vinifera (grape)	2n = 38	500 Mbp	29,971
Zea mays (maize)	2n = 20	2.4 Gbp	39,591

Source: Ensembl Plants release 44, July 2019, (EMBL-EBI), http://plants.ensembl.org/species.html.

In general, genome size does not correlate with the organism's structural complexity (Peterson and Arick II, 2018) and the number of its protein-coding genes (see Table 2.1). For example, banana has a bigger genome at 473 Mbp compared with rice (427 Mbp) but rice has more coding genes. In contrast, maize has significantly larger genome than the banana and rice at 2.4 Gbp but its gene number is less than rice and just slightly more than banana. This non-correlation between genome size and gene number is called the C-value paradox. Several plant genomes have already been sequenced and from the published genomes, the average number of protein-coding genes in plants is approximately 32,000. On the other hand, genome size is positively correlated with the quantity of repetitive sequences (Michael, 2014).

Larger genomes are more costly to sequence. In fact, only a few genomes larger than 10 Gbp have been sequenced, which include wheat, *Ginko* L., and *Pinus* L. (Li and Harkess, 2018).

2.1.2. Chromosome Number and Structure

It was discussed in the previous chapter that DNA is packaged into discrete chromosomes which can be isolated, stained, and arranged into a karyotype or karyogram (see Figure 2.2). While chromosome number and size are generally fixed for each species, these vary across organisms (see Table 2.1). Chromosome number is designated with the letter n for haploid, 2n for diploid, and so on. A true diploid has somatic cells with 2C = 2n = 2x and gametes that are 1C = n = 1X. A haploid chromosome number corresponds to a DNA content of 1C and so when chromosome number becomes 2n, correspondingly, the DNA content also becomes 2C. Since chromosomes were first seen under the microscope, it is counted as one when it appears as a single unit. Therefore, when the DNA content of the chromosomes in a diploid cell is replicated before cell division, the DNA content becomes 4C but the chromosome number is still 2n because there is no change in the number of units seen under the microscope. However, the chromosomes are also replicated so that it is now made up of two sister chromatids that are joined together at the centromere. Once these sister chromatids are separated during cell division, each one is now counted as a distinct chromosome (Peterson and Arick II, 2018).

Among plants, the highest chromosome number recorded is that of *Sedum suaveolens* (stonecrop) which is 2n = 640 while the lowest is 2n = 4 observed in the flowering plant *Haplopappus gracilis* (Cullis, 2004). Analysis of sequenced plant genomes indicates that larger genomes have fewer chromosomes while smaller genomes have more chromosomes (Michael, 2014). On the other hand, higher chromosome number is correlated with reduced chromosome size. Variation in chromosome structure is also observed among various plant species. While most have the usual chromosomes with a single centromere, i.e., monocentric, some have holocentric chromosomes where the kinetochores are found along the length of the chromosome arm (see Figure 2.3). The type of chromosome structure affects chromosome behavior during meiosis. Specifically, a holocentric chromosome may be fragmented into several segments with their own kinetochore. This is not lethal among plants. In fact, this phenomenon confers plants with strong resistance to radiation damage (Cullis, 2004). Additionally, since the segments with kinetochore are properly segregated during chromosome separation, this will result into increased chromosome number. Plants can generally tolerate an increase in their ploidy level as opposed to humans and animals wherein such changes prove to be deleterious (Cullis, 2004; Mandrioli and Manicardi, 2012).

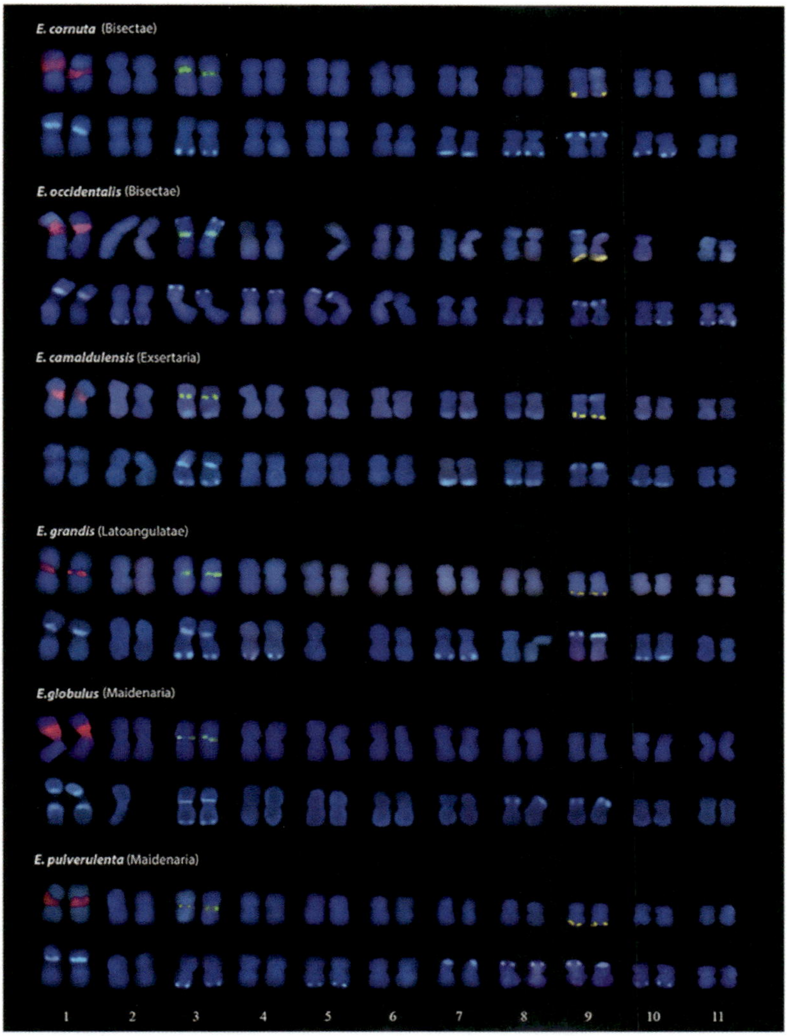

Figure 2.2: *Comparison of the karyotype of six species of Eucalyptus shows that these have 2n = 22 chromosome number. Most of the chromosomes are metacentric (i.e., centromere at the middle) with some submetacentric types.*

Source: Ribeiro et al. (2016).

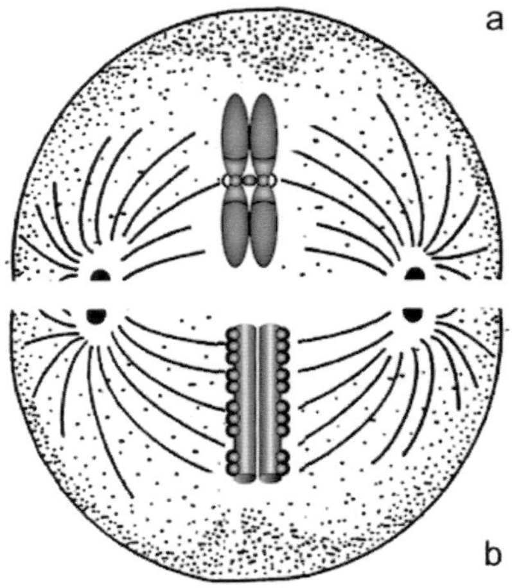

Figure 2.3: *(a) A monocentric chromosome has a single centromere with kinetochore where the spindle fibers eventually attach themselves during cell division. (b) A holocentric chromosome has several kinetochores and so when the spindle fibers shorten, the chromosome may be split into several segments.*

Source: Mandrioli and Manicardi (2012).

Many plants are polyploids. In some cases, one genus may have member species with variable ploidy levels (see Figure 2.4). Therefore, increase in ploidy level is one mechanism for the formation of new species in plants. This maybe a result of spontaneous or induced chromosome doubling. Polyploids may have several copies of the same genome. These are called autopolyploids. For example, an autotetraploid (4X = 4n) has four identical genomes which means it has four sets of homologous chromosomes which may form quadrivalents during meiotic synapsis. On the other hand, two plants belonging to different species may be crossed to form a hybrid with different genomes and when these genomes are doubled, allopolyploids are formed (see Figure 2.4). This type of polyploids may function as diploids. An allotetraploid, for instance, has four genomes but comes from two different sources so that there are just two sets of homologous chromosomes (4X = 2n). Thus, only bivalents are formed during meiosis. Plant breeders may use chemical agents, such as colchicine, to prevent chromosome segregation during cell division resulting in a nucleus with twice the number of chromosomes. Polyploid plant varieties are preferred since these are larger in size with larger parts, such as flowers or fruits, or with higher yield compared to their diploid relatives (Griffiths, 1999). Another advantage conferred by polyploidy, particularly triploids, is infertility which means the plants produce fleshly, but seedless fruits such as in cultivated banana, grape, and watermelon (Pearson, 2001).

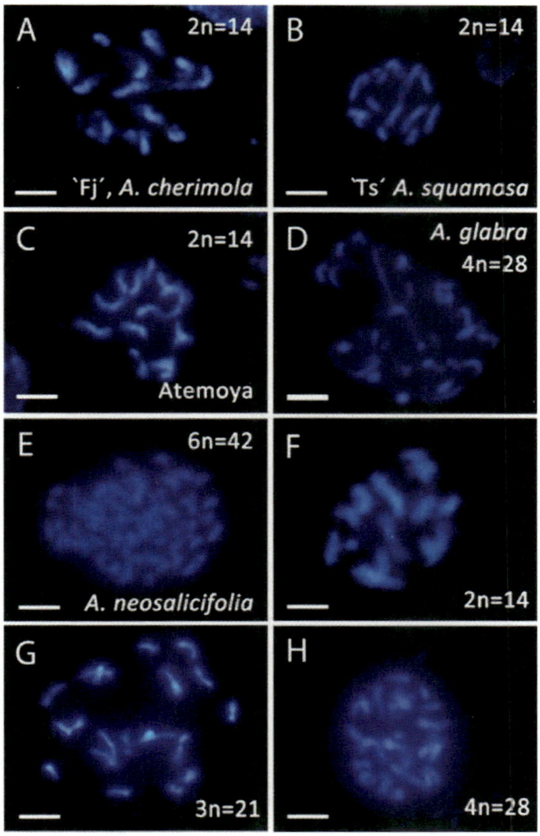

Figure 2.4: *Several species of the fruit-bearing tree Annona, which has a base genome number of n = X = 7, exhibit different ploidy levels: (A) A. cherimola, cv. Fj is diploid, 2n = 14; (B) A. squamosa, Ts is diploid, 2n = 14; (C) atemoya (hybrid of A. cherimolax A. squamosa) is diploid, 2n = 14; (D) A. glabra is a tetraploid, 4X = 28; (E) A. neosalicifolia is a hexaploid, 6X = 42. Interspecific crosses between A. cherimola and A. squamosa may also produce hybrids of varying ploidy levels: (F) a diploid hybrid, 2n = 14; (G) a triploid hybrid, 3n = 21; (H) a tetraploid hybrid, 4n = 28.*

Source: Martin, Viruel, Lora, and Hormaza (2019).

Sequencing polyploids is more complicated due to the presence of more genomes. Thus, it is recommended to sequence a diploid representative of the taxonomic class of interest, unless the polyploids are the target individuals for sequencing (Li and Harkess, 2018).

2.2. ACQUIRING GENOME DATA

As sequencing technology develops, the cost of the procedure also decreases. Thus, sequencing non-model plants become more feasible. This section describes the general steps for sequencing plant genomes. The previous chapter points out that a plant has three distinct genomes, the nuclear, mitochondrial and chloroplast genomes. Nevertheless, when one says that a certain plant's genome is sequenced, it is assumed that one is referring to the nuclear genome.

2.2.1. DNA Extraction

The quality of the final genome sequence depends on the quantity, purity, and integrity of the extracted DNA. This can be challenging in certain plant species due to the presence of a lot of secondary metabolites. DNA may be extracted using the conventional nuclei preparation protocols or DNA extraction kits (Li and Harkess, 2018). Pre-washing with sorbitol prior to the extraction procedure has been proven to remove interfering metabolites, such as polyphenols and polysaccharides (Inglis et al., 2018). After extraction, DNA integrity can be estimated and visualized by agarose gel electrophoresis where the desired result is a distinct, solid band of high molecular weight that is located near the loading well (see Figure 2.5). A quantitative measurement of total genomic DNA may be done using double-stranded DNA fluorescence-based quantification. DNA with high purity is also required because salt and protein contaminants may inhibit enzyme activity during the clone library preparation. This can be measured with an ultraviolet-visible (UV-Vis) spectrophotometer (see Figure 2.6) that can measure absorbance at 230 nm, 260 nm, and 280 nm, which are the wavelengths for measuring the optimal absorbance of salts, DNA, and proteins, respectively. Generally, pure DNA should have a 260/280 value between 1.8–2.0, and a 260/230 value between 2.0–2.2 (Li and Harkess, 2018).

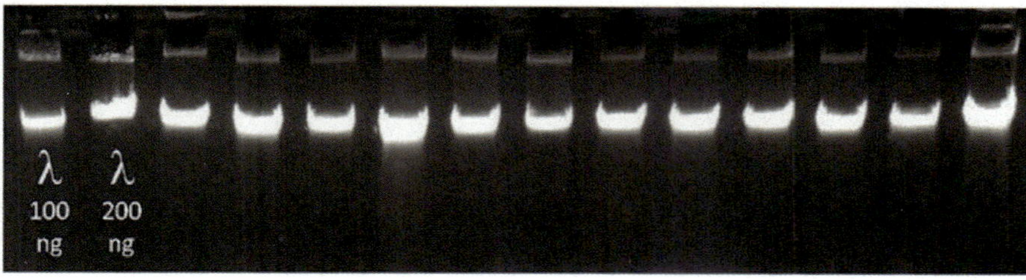

Figure 2.5. *Genomic DNA extracted from Eucalyptus grandis (lanes 3–8) visualized on agarose gel. Lanes 1 and 2 contain 100 and 200 nanograms (ng) of uncut Lambda (λ) DNA which are used as bases for estimating the quantity of the DNA in each lane.*

Source: Inglis, Pappas, Resende, and Grattapaglia (2018).

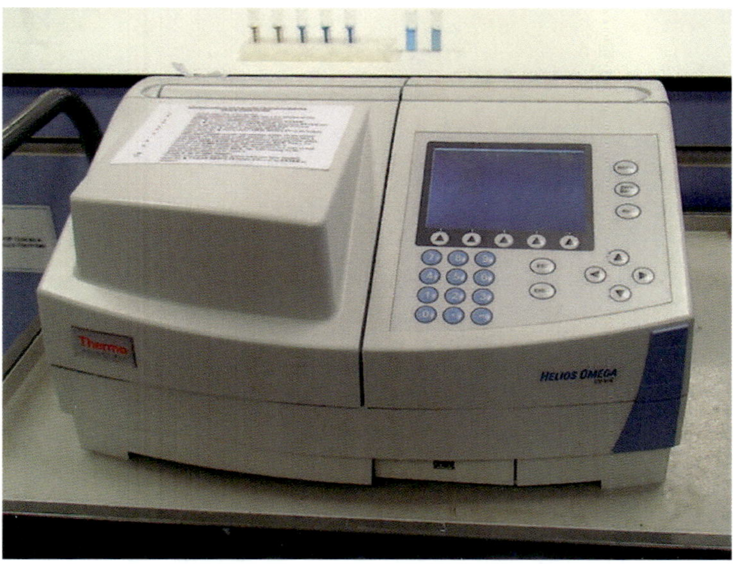

Figure 2.6. *An ultraviolet-visible (UV-Vis) spectrophotometer may be used to determine the purity of the extracted DNA.*

Source: Viv Rolfe, Creative Commons License.

2.2.2. Preparing the DNA Library

Traditionally, the extracted genomic DNA is then fragmented by restriction enzyme digestion in order to make the DNA library. The DNA fragments are then inserted into vectors (usually a plasmid), which are used to transform bacterial cells or virions. This collection of bacterial cells or virions containing recombinant vectors with DNA sequences from a certain organism is the DNA library (see Figure 2.7) (Peterson and Arick II, 2018).

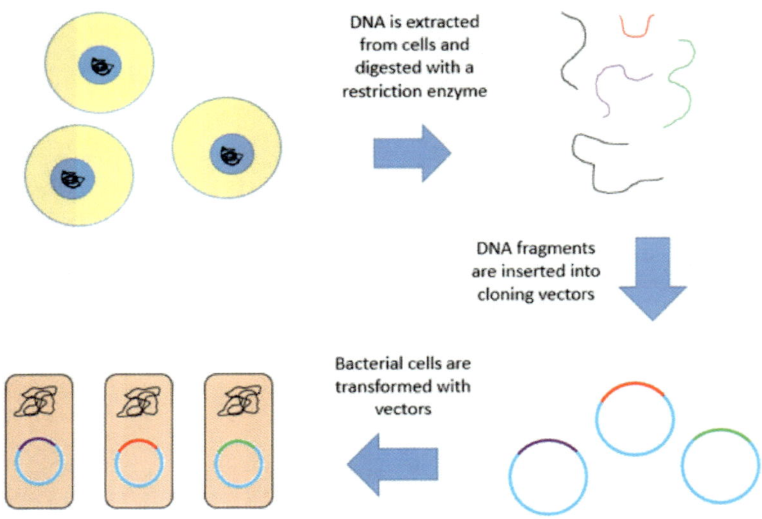

Figure 2.7: *A DNA library is prepared from the extracted genomic DNA that is eventually digested by restriction enzymes. The resulting fragments are inserted into vectors which are eventually taken up by bacterial cells.*

Source: Aluquette, Creative Commons License.

With the advent of next-generation sequencing (NGS) technology, the term DNA library simply means DNA prepared in a way that is compatible with a specific sequencing procedure. Preparing DNA for NGS depends on the platform to be used (see Figure 2.8) but it generally involves cutting the DNA into fragments of a desired length. DNA may be fragmented by physical methods, such as acoustic shearing and sonication, or enzymatic digestion (Head et al., 2014). Many NGS platforms require the attachment of oligonucleotide adapters onto the ends of the DNA fragments. Finally, these are quantified for sequencing (Peterson and Arick II, 2018; Head, 2014).

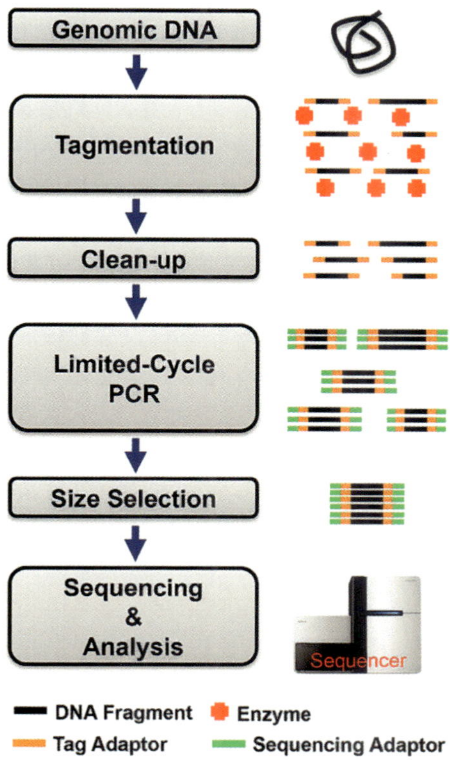

Figure 2.8: *The Nextera®* protocol for preparing a DNA library *that is compatible with Illumina sequencers. After extracting the genomic DNA, it undergoes tagmentation, which involves the simultaneous transposon-mediated cutting of the double-stranded DNA and tagging it with a universal overhang. The tagged DNA fragment library is then purified to remove unwanted compounds in the reaction solution. Sequencing adapters are added to the fragments by limited-cycle polymerase chain reaction (PCR). Finally, the desired size of the DNA library is selected for sequencing and analysis.*

Source: Kim et al. (2013).

There are a variety of methods for quantifying DNA library used for NGS, such as measuring UV absorbance via a spectrophotometer, employing intercalating dyes, utilizing 5' hydrolysis probes with real-time quantitative PCR (qPCR), or droplet digital emulsion PCR. Most of these methods require large amounts of DNA library preparations (in micrograms). In some cases, this amount is obtained by performing several rounds of amplification using PCR (Robin et al., 2016).

2.2.3. Determining the DNA Sequence

Sequencing is the process where the identity and order of the nucleotides that make up the plant genome are identified. Generally, the amount of DNA required for sequencing is approximately 1000-fold less than what is needed to quantitate the DNA library (Robin et al., 2016). In a genome sequencing project, the genome is fragmented into millions of small pieces of single-stranded DNA which are sequenced producing the so-called reads. These range in length from 20 to thousands of base pairs. These may be single reads generated by sequencing the DNA from only one end (see Figure 2.9B). In contrast, paired-end reads are produced by sequencing the DNA fragment from its two ends (Gladman, 2019). Having a second read allows a reliable mapping of regions in the genome that contain repetitive sequences or repeats (see Figure 2.9A). In contrast, single-end reads may be mapped multiple times in the repeat-rich regions of the genome presenting a challenge in making an accurate genome assembly (Grehl et al., 2018) (see Figure 2.9B).

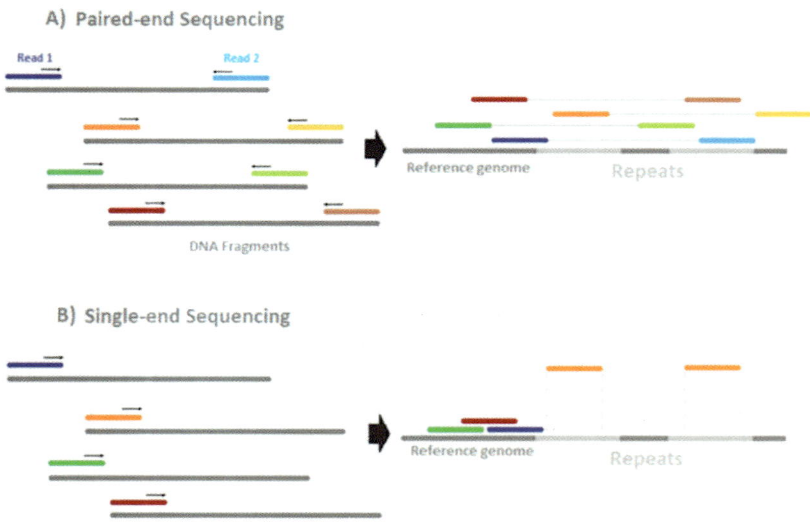

Figure 2.9: *(A) DNA fragments are sequenced by synthesis from both sides producing two reads called paired-end reads. These can potentially map deeper into the repeat-rich regions of the genome. (B) Single-end sequencing produces reads from only one side of the fragment. These may be mapped multiple times in repeat-rich genomes.*

Source: Grehl et al. (2018).

2.2.3.1. Sanger Sequencing

An early method of determining DNA sequence was developed by Frederick Sanger in 1975. Hence, it was named Sanger sequencing method. It is one of the so-called first-generation sequencing methods along with the Maxam-Gilbert sequencing method. Sanger sequencing was more widely used than Maxam-Gilbert method (Peterson and Arick II, 2018). This uses the DNA to be sequenced as a template for DNA synthesis, which is catalyzed by DNA polymerase. It utilizes a DNA primer to initiate the synthesis. Four DNA synthesis reactions are prepared, which contain normal A, G, C, and T deoxynucleotide triphosphates (dNTPs). These are the building blocks that are joined together to form a DNA strand. In addition, each reaction has a low amount of one of four dideoxynucleotide triphosphates (ddNTPs), i.e., ddATP, ddGTP, ddCTP, or ddTTP. Each of these is labeled with a different fluorescent dye. When a ddNTP is incorporated into the strand during the reaction, synthesis is terminated because it lacks a 3' hydroxyl group which is needed to form the link with the next nucleotide (Scitable by Nature Education, 2014) (see Figure 2.10). For this reason, the Sanger sequencing is also called the chain termination method. At the end of each reaction, DNA fragments of various lengths are generated, all of them ending with a labeled ddNTP (Merck, 2019).

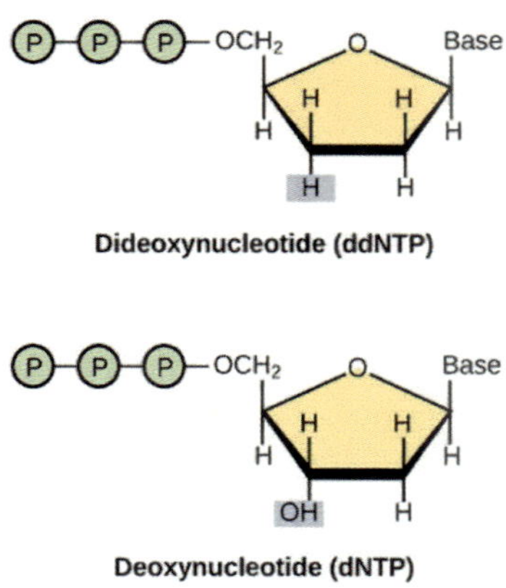

Figure 2.10: *A dideoxynucleotide (ddNTP) is similar in structure to a deoxynucleotide (dNTP) except for its lack of 3' hydroxyl (OH) group (shaded light blue). The 3' OH group is needed in forming a bond with the next nucleotide during DNA replication. For this reason, elongation of DNA stops when a ddNTP is incorporated in the DNA strand being synthesized.*

Source: OpenStax CNX, Creative Commons License.

In the early sequencing efforts, the DNA fragments produced from the four reactions are individually loaded into four lanes (designated as A, G, C, and T which correspond to the ddNTP content of each reaction) of a polyacrylamide gel and are then separated by electrophoresis. During this process, the negatively-charged DNA fragments move across the gel toward the positive electrode. The shorter fragments will move faster and migrate farther on the gel than the longer ones. This results to a gel with a series of bands (representing the DNA fragments) at various distances from the loading well with the lowest band representing the shortest fragment and the highest band the longest (see Figure 2.11). The sequence is then read from bottom to the top of the gel. For example, if the lowest band appears on the A reaction lane, then the first nucleotide in the sequence is A. If the next band from the bottom appears in the C lane, then the second nucleotide is C, and so on (Scitable by Nature Education, 2014).

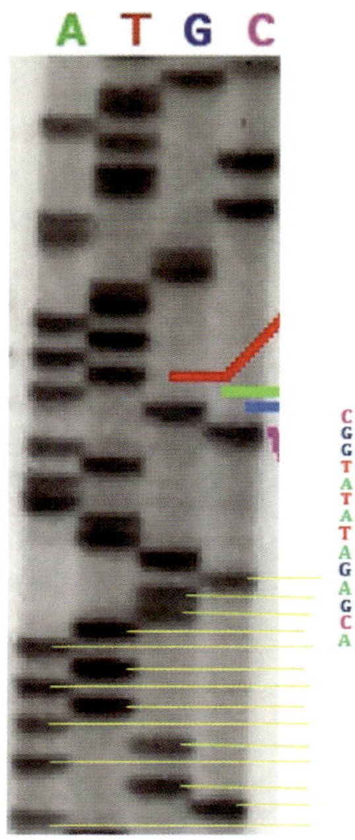

Figure 2.11: *The DNA products of the four chain termination reactions via dideoxynucleotide (ddNTP) are loaded into separate lanes of a polyacrylamide gel. The ladder-like bands represent the DNA fragments arranged from shortest (bottom) to longest (top). Thus, the sequence of the entire DNA is read from bottom to top.*

Source: John Schmidt, Creative Commons License.

Eventually, capillary electrophoresis replaced polyacrylamide gels. This simplified and automated the procedure and thus, increased the output. In this step, the reaction products are injected into a capillary, which is a long glass tube filled with gel polymer (see Figure 2.12). If this method of electrophoresis is employed, the four ddNTPs may be put together with the template in one mix. In this case, the reaction products are placed in a single capillary. On the other hand, if each ddNTP has its own reaction set-up, then four capillaries are needed to separate the four reaction products (Merck, 2019). Similar to polyacrylamide gel electrophoresis, the fragments are also separated based on their size or molecular weight. Just before reaching the positive electrode at the end of the capillary, the fluorescently labeled DNA fragments cross the path of a laser beam, which activates and causes the dye on the fragment ends to fluoresce. Expectedly, the smallest fragment will move across the laser beam first, followed by the next smallest, and so on (Schoales, 2015; Khan Academy, 2016).

Since the four types of ddNTP are each labeled with a different fluorescent dye, they emit fluorescence at different wavelengths. These are detected by an optical device and recorded in a chromatogram or electrophoretogram which shows the fluorescence peak of each nucleotide (see Figure 2.12). Certain software converts the fluorescence signals to digital data, translating a certain signal to a corresponding nucleotide so that the DNA sequence is eventually determined (Scholaes, 2015; Applied Biosystems, 2009; Khan Academy, 2016).

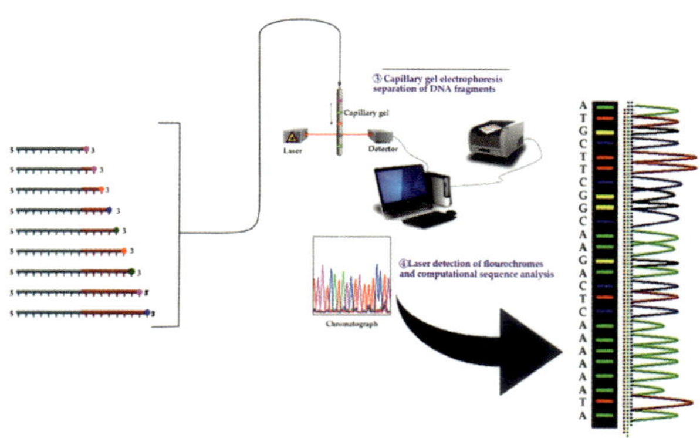

Figure 2.12: *Capillary electrophoresis separates the DNA fragments from the Sanger reactions based on molecular weight. As each fragment moves past a laser, the fluorescently labeled dideoxynucleotide (ddNTP) at the end of the fragment gets activated. Each type of ddNTP emits fluorescence at a different wavelength. The resulting chromatogram is translated by a computer software into a DNA sequence.*

Source: Left image: Estevesj, Creative Commons License; Right image: Abizar, Creative Commons License; Arrow: SimpleIcon, Creative Commons License.

The human genome and the *A. thaliana* genome were sequenced using the Sanger sequencing method. It accurately reads long segments up to 800–1000 bp. However, it is low throughput and so it takes a long time to finish sequencing a relatively large genome. It is also costly at a price of US$500,000–1,000,000/Gbp (Li et al., 2018). As an example, the *A. thaliana* genome took 10 years to finish at a cost of about US$ 100 million (Rhoads and Au, 2015; Li and Harkess, 2018).

2.2.3.2. Next-generation sequencing (NGS)

DNA sequencing technologies have developed rapidly that the *A. thaliana* genome can be sequenced within a week at a cost of under US$1000 (Li and Harkess, 2018). These post-Sanger sequencing technologies are collectively called NGS. These are high throughput enabling the sequencing of genomes at a faster pace and cheaper cost. As a result, the number of sequenced plant genomes has increased to about 200 species (Li et al., 2018).

There are three major advantages of NGS over first-generation sequencing methods. The first one is that these do not require bacterial cloning. Instead, DNA libraries are prepared in a cell free system. Secondly, millions of sequencing reactions can be processed in parallel and simultaneously. Finally, bases are detected in a cyclical and parallel manner (Park and Kim, 2016).

NGS makes it possible to sequence multiple cultivars of the same plant species. This was done in 360 wild and cultivated tomato accessions, which lead to the discovery of two potential quantitative trait loci (QTL) that increased fruit size 100 times during the course of domesticating tomato. Resequencing of 3,000 rice accessions sourced from 89 countries was also conducted. Results show significant genome diversity among the accessions which can possibly lead to the discovery of novel alleles. These data are useful in accelerating rice breeding and improvement (Li et al., 2018).

2.2.3.2.1. Second-generation sequencing

Second-generation sequencing techniques convert the starting DNA fragments into double-stranded molecules that are flanked by adapters, which are oligonucleotides that enable the binding of target DNA to a substrate, which is either a bead or a flowcell (see Figure 2.13). Amplification by PCR ensues to generate thousands of DNA on the substrate, whose sequence is then determined by detecting light, fluorescence, or pH (Knief, 2014). This amplification step increases signal intensity which may reduce sequencing error. Multiple sequencing reactions can occur and be monitored at once. However, not all of the reactions are 100% efficient. For instance, some of the DNA fragments may have missed a base incorporation or ligation causing these to be out of synch with those who have. This asynchronization becomes amplified with each additional cycle creating noise during sequencing. This is the major factor why second-generation sequencers initially generated short sequence reads, about 300–500 bp, but these were increased by improvements in the technology (Peterson and Arick II, 2018).

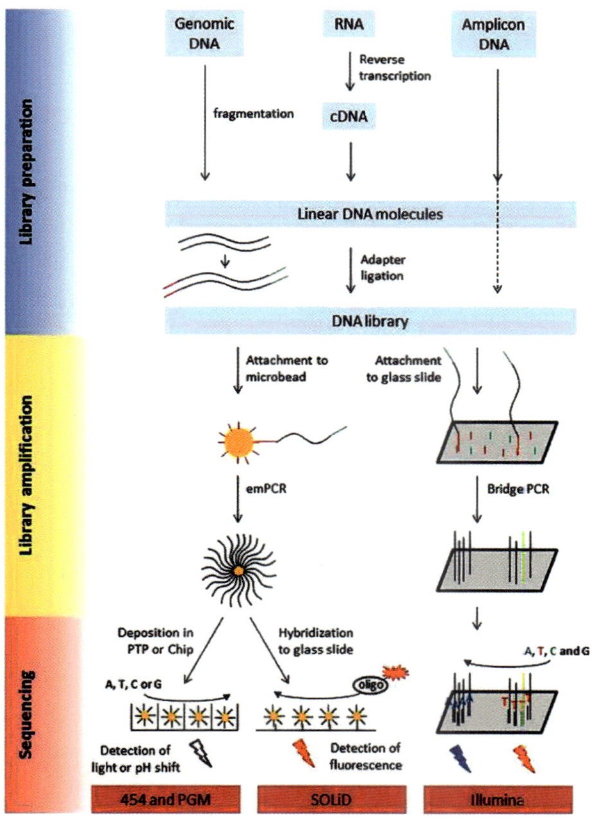

Figure 2.13. *The commonly used second-generation sequencing platforms share certain similarities. They prepare their DNA library by making their starting DNA material double-stranded and ligating it to adapters. These are then amplified into thousands of copies by PCR prior to sequencing. The sequenc-*

ing method varies among the platforms in terms of the data that is collected, whether it is fluorescence, light, or pH.

Source: Knief (2014).

2.2.3.2.1.1. 454 Pyrosequencing

The 454 pyrosequencing method is the first commercial high throughput sequencing platform that was developed (Peterson and Arick II, 2018). In this method, the sequence is read as the DNA strand is being extended. This eliminates the need for electrophoresis to identify the nucleotide sequence (Farrar, 2019). Compared to other second-generation sequencing methods, it produces much lesser reads at 1 million but generates longer reads with an average length of 700 bp (Knief, 2014).

This method depends on the chemistry of DNA replication. Target DNA are sheared up to 1 kb in order to be used as templates. These are then ligated to short DNA segments called A and B adapters and those flanked by an adapter on one end and a B adapter on the opposite end are isolated. These isolated DNA are rendered single-stranded and are mixed with an excess quantity of beads that contain thousands of capture probes that are complementary to adapter A. The low amount of DNA relative to the beads ensures that only one single stranded DNA molecule binds to a bead.

The bound target DNA is amplified by emulsion PCR (emPCR) (see Figure 2.14). PCR reagents and oil are added to the beads and these are agitated to create an emulsion. The size of the bead promotes the formation of an aqueous layer of PCR reagents around each bead. The hydrophobicity of the oil keeps the beads separated from each other. After emPCR, thousands of identical DNA molecules are attached to the surface of each bead. Then the emulsion is broken and only those DNA strands synthesized through the elongation of capture probes remain attached to the beads forming bead colonies or polonies. These beads are allowed to fall into a slide with microfabricated wells containing DNA polymerase and sequencing reagents, with one bead per well.

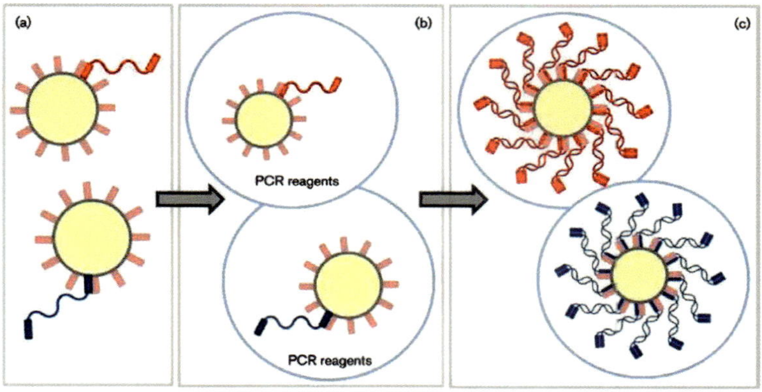

Figure 2.14: *Emulsion PCR (a) use adapters (red and blue bars) to capture a single stranded DNA template onto the microbeads via hybridization with the probe (pink bar). (b) The microbeads are then incorporated into an emulsion of PCR reagents. (c) After amplification, each bead is coated with thousands of identical DNA strands.*

Source: Radford et al. (2012).

The sequencing reaction is initiated by the annealing of a primer to the B adapter in the sequences. The slide is flooded with one of the four types of dNTPs. If it is the next nucleotide in the sequence, then it is successfully added to the primer end (see Figure 2.15). After addition, the diphosphate or pyrophosphate (PPi) group of the dNTP is released and used in a reaction involving luciferase to produce light. This reaction is called chemiluminescence. The location of the produced light is captured and used to determine which beads the nucleotide is added to. The remaining dNTP mix is washed off and the next dNTP mix is added, and the process is repeated, cycling through the four dNTPs. The signal pattern and density for each dNTP wash are presented as a graph. From this data, the nucleotide sequence can be computationally determined using a software (Peterson and Arick II, 2018; European Bioinformatics Institute, 2019a).

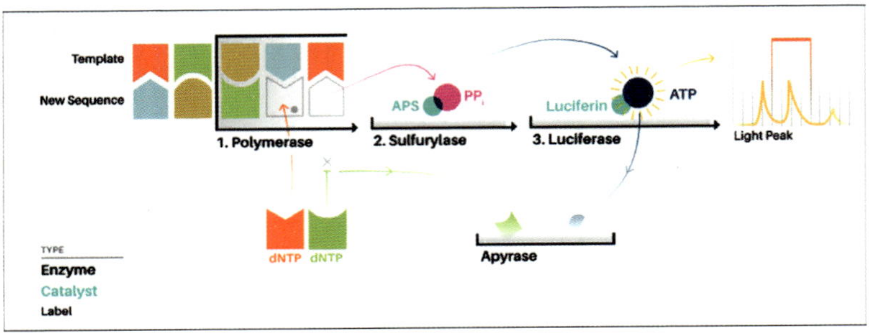

Figure 2.15: *In 454 pyrosequencing, the dNTP added into the DNA being synthesized releases its pyrophosphate group (PPi), which initiates a chemiluminescent reaction. The light emitted is detected and the resulting density and pattern of signal emission is used to determine the sequence of nucleotides in the synthesized DNA.*

Source: Jacopo Pompilii, Density Design Research Lab, Creative Commons License.

Being the first NGS in the market made 454 pyrosequencing platforms commercially successful at first. However, emPCR preparation for this method is time consuming, technically challenging, and expensive. Additionally, this method has difficulty in distinguishing a sequence of longer than six identical nucleotides (e.g., CCCCCCCC) (Peterson and Arick II, 2018; Farrar, 2019).

2.2.3.2.1.2. Illumina Sequencing

Illumina platforms were not the first second-generation sequencing technology, but these have been widely used for genome sequencing. These can generate massive sequencing data at a low cost and with a low error rate. These produce 22–25 million reads. However, these have relatively short read length, a maximum of 150 bp for HiSeq and 300 bp for MiSeq. Moreover, the library insert size cannot be longer than 800 nucleotides so it can only provide short-range data. Therefore, these are ideal for small genomes (Li and Harkess, 2018; Knief, 2014).

Illumina sequencing uses reversible dye terminators that enable the identification of each nucleotide as it is incorporated into the DNA strand during replication (see Figure 2.16). The first step is to break up the DNA molecule and denature them into fragments of approximately 200–600 bp. Adapters are then attached to both ends of these fragments (Raghavendra and Pullaiah, 2018). Primers are placed on a slide or flowcell and the DNA fragments are washed across it allowing them to hybridize with the primers they are complementary with. The attached fragments bend over since the adaptor in the free end also hybridizes with another primer and these form bridges. The DNA molecules that failed to attach to the primers are washed away. This is followed by DNA amplification via several cycles of bridge PCR (Zhou and Li, 2015). Primers, unlabeled dNTPs, and DNA polymerase are added to replicate the DNA strands attached to the flowcell. The resulting double stranded DNA are separated into single strands. The synthesized reverse strands are washed off generating millions of dense clusters made up of identical DNA sequences that serve as template for the replication step (Raghavendra and Pullaiah, 2018; Peterson and Arick II, 2018).

The four types of reversible terminating dNTPs are fluorescently labeled, each type with a different color, and its 3' end attached to a blocking or terminating group. These, along with primers, are added to the clusters where the primer attaches to the DNA template and the DNA polymerase adds any one of the four dNTPs to the primer as it lengthens the new DNA strand. Once a nucleotide has been added, no other dNTP can be added to the DNA being synthesized until the blocking group is removed. The excess dNTPS which are not incorporated in the new strand are then washed off. After the addition of each nucleotide, a laser is applied to remove the blocking group and to activate the fluorescent label of the nucleotide. The resulting fluorescence is detected by a camera and recorded on a computer. Since the blocking group has been removed, a new fluorescently labeled dNTP can now be added to initiate the next cycle. The process is repeated until the entire DNA molecule is sequenced (Zhou and Li, 2015; Raghavendra and Pullaiah, 2018).

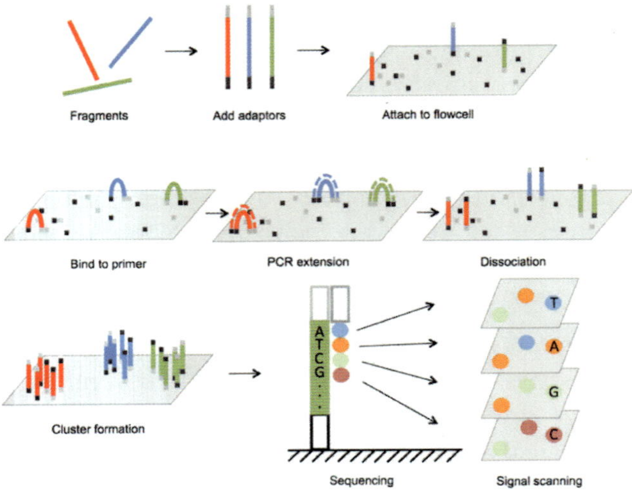

Figure 2.16: *In the Illumina sequencing method, adaptors are ligated to both ends of DNA fragments. These are then washed across a flowcell containing primers that are complementary with the adaptors so that both ends of the fragments hybridize with the primers to form bridges. These single stranded bridges are made double stranded and amplified through PCR. The double-stranded DNA products dissociate to form a cluster of identical single stranded DNA sequences, which will serve as templates for the sequencing reactions. In each sequencing cycle, a fluorescently labeled nucleotide is added to the growing strand which emits a signal that is detected and recorded by an optic scanner. The sequencing terminator in the nucleotide is removed and the next sequencing cycle starts.*

Source: Lu, Shen, Warren, and Walter (2016).

2.2.3.2.1.3. Ion Torrent Sequencing

Ion torrent sequencing deviates from other second-generation methods by not using either fluorescence or chemiluminescence to determine nucleotide identity and position. Instead, it detects base sequence from pH changes around a bead colony, which result from the H^+ released as a by-product of phosphodiester bond formation during DNA synthesis. Thus, it does not require sophisticated optics system and so it is relatively cheaper than the other methods. The sequencing process is also very fast and may take just two hours. On the other hand, it also has difficulty with segments made up of a series of identical nucleotides. The read length of this method ranges from 100 to 400 bp with 0.4 to 5.5 million reads (Peterson and Arick II, 2018; Knief, 2014).

Just like in the other second-generation sequencers, the input DNA is fragmented to about 200 bp. Adaptors are attached to both ends of these fragments and one DNA fragment is placed onto a bead. The fragment is then amplified by emPCR and then one bead is placed per well of a flowcell for the subsequent DNA synthesis step (European Bioinformatics Institute, 2019b). Underneath each of these wells are pH sensors that can detect even the smallest changes in pH (Snipacademy, 2019). Similar to 454 pyrosequencing method, the flowcell is washed with a single type of dNTP at a time, along with buffers and DNA polymerase. Upon the addition of each dNTP to the elongating DNA strand, H^+ is released and will decrease the pH (see Figure 2.17A). The change in pH is used to determine which base and how many thereof was added to the new strand (see Figure 2.17B). The remaining dNTPs are washed away, and the process is repeated, using a different type of dNTP until all four types are utilized (European Bioinformatics Institute, 2019b).

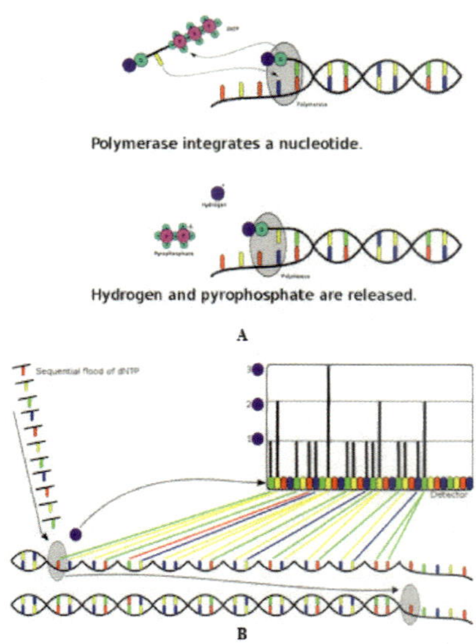

Figure 2.17: *(A) During the DNA synthesis step of the ion torrent sequencing method, H^+ and pyrophosphate are released when the dNTP is incorporated into the growing DNA strand. (B) Released H^+ changes the pH which is then detected by an ion sensor and translated into DNA sequence.*

Source: David Tack, Creative Commons License.

2.2.3.2.2. Third-Generation Sequencing

Plant genomes are characterized by having repetitive regions so that the short reads generated by second-generation sequencing cannot fully span over these regions. This results to incomplete genomes with missing sequences (Li et al., 2018). Third-generation sequencing technologies provide much longer reads of more than 10 kbp which can close these gaps. However, these produce lower quality reads and lower throughputs compared to second-generation techniques. These are also more expensive, so these are ideal for small to moderate-sized genome with less than 1 Gbp. For larger genomes, a hybrid approach that combines short reads from second-generation sequencing and long reads from third-generation methods may be employed in order to cut back on the cost as well as reduce the error rates (Peterson and Arick II, 2018; Li and Harkess, 2018; Rhoads and Au, 2015).

One main deviation of third-generation sequencing from the second-generation techniques is that the former no longer has amplification step but sequence directly from the template DNA making it useful for *de novo* sequencing of plant genomes (Park and Kim, 2016).

2.2.3.2.2.1. PacBio Sequencing

The PacBio sequencing platform is named after Pacific Biosciences. This method is also called single molecule, real time (SMRT) sequencing. Similar to the second-generation sequencing methods, it depends on the principle of DNA replication. This has been employed to sequence a number of plant genomes including those of *Zea mays* with 2300 Mbp and *Helianthus annuus* with 3000 Mbp (Li et al., 2018).

The template is produced by attaching hairpin adaptors to a target double-stranded DNA. This forms a single-stranded circular DNA called SMRTbell (see Figure 2.18). This is then loaded into a chip called a SMRT cell, a perforated metal film on a glass surface with densely arranged nano-chambers or sequencing units called zero-mode waveguide detectors (ZMWs), which provides the smallest available volume for light detection. Each cell has 150,000 ZMWs allowing parallelization of the sequencing reactions (Rhoads and Au, 2015; Knief, 2014).

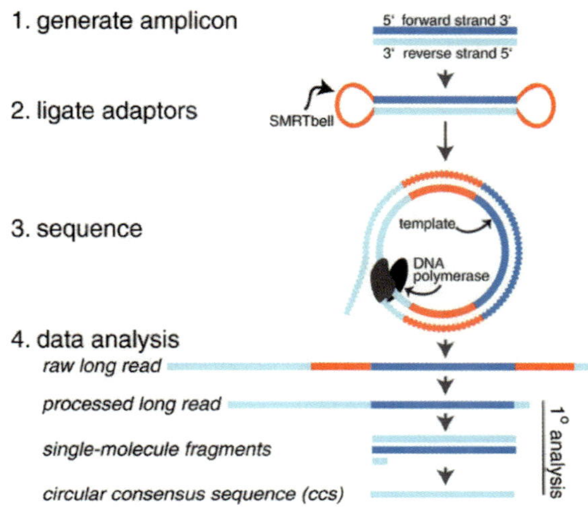

Figure 2.18: *The SMRTbell template is formed by ligating hairpin adaptors (red) to the end of a double-stranded DNA molecule (blue and light blue). This forms a circular, single-stranded DNA. The DNA polymerase (black) adds fluorescently labeled dNTPs into the elongating strand (outer strand), which is eventually sequenced.*

Source: Fichot and Norman (2013).

A single DNA polymerase molecule is attached on the bottom of each ZMW. The four types of dNTPs are also loaded in the ZMW, each with a distinct fluorescent label. When a SMRTbell is loaded to the cell, it diffuses into a ZMW where the DNA polymerase incorporates the fluorescently labeled dNTPs into a growing DNA strand. The fluorescent label is cleaved off during DNA synthesis and then the DNA polymerase moves to the next position. This occurs continuously without the intermittent washing steps which are done in second-generation techniques. This significantly speeds up the process. The fluorescence emitted by each dNTP is detected by the ZMW and so the replication processes in the ZMWs of a SMRT cell are recorded as a movie of light pulses. A confocal imaging system records the fluorescent signals from each ZMW and from these, the sequence is deduced. Approximately 35,000–75,000 of the ZMWs produces a read lasting 0.5–4 hours which correspond to 0.5–1 Gbp of sequence. Therefore, while PacBio sequencing generates long reads, it has a low throughput unlike the Illumina sequencers (Knief, 2014; Rhoads and Au, 2015).

After the DNA polymerase replicates the target DNA, it can go on and replicate the adapter and then the other strand (see Figure 2.18). This is possible because the SMRTbell is circular. The product is a continuous long read (CLR) which contains adaptor sequences as well as the forward and reverse strands. This undergoes processing where the adaptor sequences are bioinformatically removed and the CLR is split into subreads or single molecule fragments. These multiple subreads are aligned to generate the circular consensus sequence (CCS) (Rhoads and Au, 2015; Fichot and Norman, 2013). This CSS strategy improves the sequence read quality (Knief, 2014).

2.2.3.2.2.2. Oxford Nanopore MinION Sequencing

The Oxford Nanopore MinION sequencing platform is similar to PacBio sequencing in terms of producing long read lengths (but shorter than that of PacBio), its low throughput, and relatively high error rate compared to the second-generation sequencing techniques. However, its edge over the latter is its portability since it is packaged into a USB-sized device and it only requires a moderately modern laptop. Library preparation can be completed fast taking only 10 minutes. Because of this, MinION has been used in remote research stations such as the International Space Station and in the arctic. This makes it possible to perform DNA sequencing real-time in the field (Li and Harkess, 2018). In terms of plant genome sequencing, this has been used to assemble *de novo* the genome of a *Solanum pennellii* accession with a size of 1.2 Gbp. The genome of *A. thaliana* was also sequenced using the MinION sequencer covering 100% of the non-repetitive genome (i.e., 119 Mbp), the centromeres, and the ribosomal DNA (Li et al., 2018).

A nanopore is a nano-scale hole, which can be created by proteins that puncture electrically-resistant polymer membranes. Each membrane is supported by a micro scaffold that is attached to a sensory array chip. Oxford Nanopore passes anionic current through nanopores and measures the changes in current as biological molecules pass through or near the nanopore (Oxford Nanopore Technologies, 2018). In the case of DNA sequencing, a strand of DNA that serves as template for replication is passed through a nanopore (see Figure 2.19). Each nanopore has a DNA polymerase positioned near the pore entrance. On the other hand, modified dNTPs are used to make the new strand. Specifically, these carry chemical tags which are distinct for each type of nucleotide. As the DNA polymerase adds a dNTP into the growing strand, the base specific tag is captured by the nanopore (Roche, 2019). This partially blocks the channel in the sensor, which interrupts the flow of current. The four distinct tags of the four types of dNTPs block the channel differently giving a different amplitude and duration of the blockade (Roche, 2019; Bite Size Bio, 2017). This change in current is detected by the sensor and measured by application-specific integrated circuit (ASIC) to determine the DNA sequence (Oxford Nanopore Technologies, 2018).

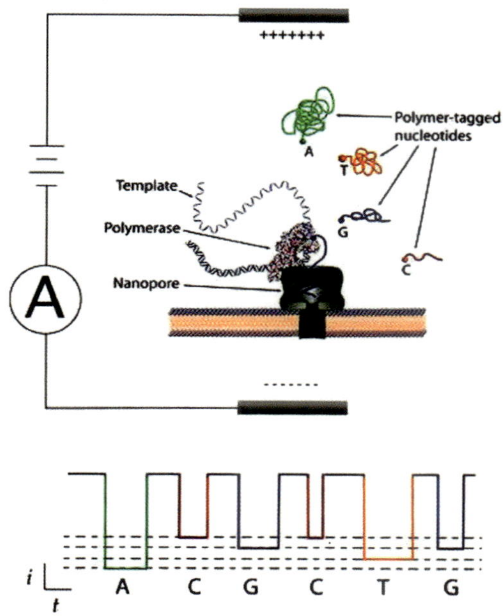

Figure 2.19: *In nanopore sequencing, ionized fluid is forced through the nanopore of an artificial membrane. A DNA polymerase attached near the nanopore opening performs DNA replication by complementing a DNA template using dNTPs that are modified with polymers. Each polymer tag is unique, in terms of chemical structure, for each type of dNTP. As each dNTP is incorporated into the growing strand, it passes through the nanopore and produces a unique ionic current signature. This can be used to determine the identity of the nucleotide.*

Source: National Institute of Standards and Technology, Public Domain.

Table 2.2 compares the different sequencing methods that were discussed in this section in terms of how the DNA library is amplified and how the nucleotide identity is determined. Moreover, it summarizes the output quality, speed of processing, and cost of each method.

Table 2.2: *Specification and Performance Comparison of Various Generations of Sequencing Platforms*

Method	Generation	Library Amplification	Signal Detection Method	Read Length (bp)	Single Pass Error Rate (%)	No. of Reads per Run	Time per Run (Hours)	Cost per Million Bases (US$)
Sanger ABI 3730×1	1st	PCR	Optical detection of fluorescence	600–1000	0.001	96	0.5–3	500–1000
454 (Roche) GS FLX+	2nd	Emulsion PCR on microbeads	Optical detection of light	700 (ave.) 1,000 (max.)	1	1 million	23	8.57
Illumina HiSeq 2500 (High Output)	2nd	Bridge PCR on flow cell	Optical detection of fluorescence	2 × 125	0.1	8 billion (paired)	7–60	0.03
Illumina HiSeq 2500 (Rapid Run)	2nd	Bridge PCR on flow cell	Optical detection of fluorescence	2 × 250	0.1	1.2 billion (paired)	24–144	0.04
Ion torrent	2nd	Emulsion PCR on microbeads	Transistor-based detection of pH	200	1	82 million	2–4	0.1
PacBio RS II:P6-C4	3rd	Not applicable	Real-time optical detection of fluorescence	10,000–15,000	13	35–75 million	0.5–4	0.125
Oxford Nanopore MinION	3rd	Not applicable	Detection of current	200,000 (max.)	38	11,000–47,000	48–50	0.750

Sources: Rhoads and Au (2015); Knief (2014); Li et al. (2018).

2.2.4. Assembling the Genome

After generating sequence reads of the target DNA, the next step is to assemble these into a genome sequence (see Figure 2.20). Genome assembly is like a putting together a jigsaw puzzle since it puts back together the sequenced fragments and create a representation of the original chromosomes from which they originated. De novo assemblies assume no prior knowledge of the sequence length, layout, or composition (Gladman, 2019). In contrast, reference-based assembly utilizes a previously generated genome as a basis. The sequenced reads are aligned against the reference genome sequence and then each one is placed at its most likely position.

Theoretically, an ideal de novo assembly would involve finding the common string (genome) of all substrings (reads). There are two approaches used, the de Brujin graph (DBG) or the overlap-layout-consensus (OLC) algorithm (Li et al., 2012).

The OLC assembly approach finds the overlaps among all the sequencing reads. From these, a string graph is created to lay out the contigs. Contigs refer to the contiguous sets of overlapping sequenced DNA fragments. The reads that make up each contig is then used to infer the consensus sequence. OLC is not designed for short reads since the overlaps between short sequences could be inaccurate. Also, it is computationally impossible to calculate pair wise overlaps among billions of reads. Thus, OLC assemblers are usually utilized to assemble long reads generated through PacBio or Nanopore sequencing (Li and Harkess, 2018; Li et al., 2012).

DBG is better suited for assembling very large number of short reads. It further shreds these reads into much shorter *k*-mers (or sequences). The principle behind this is that it makes it easier to construct the connections among these shorter *k*-mers. The resulting DBG can then be traversed to derive the contigs (Li and Harkess, 2018; Li et al., 2012).

As previously mentioned, sequencing large genomes usually employ a hybrid assembly approach combining both approaches. An example of this is the genome assembly software MaSuRCA which transforms numerous short reads into fewer but longer "super-reads." This enables combining Illumina reads of varying lengths with longer reads from 454 and Sanger sequencing platforms. These are then used to turn long reads into "mega-reads." These processed reads can then be assembled via OLC. Genomes assembled with MaSuRCA include those of plants like *Aegilops tauschii* (ancestral wheat) and *Pinus taeda* (loblolly pine) (Li and Harkess, 2018; Zimin et al., 2013).

Generally, an assembly that includes 90–95% of the genome is considered to be of good quality. The unaccounted percentage is due to the presence of repeats that are difficult to sequence and arrange. Getting a fully covered genome is even more challenging in plants which have abundant repetitive regions in their genomes, Genome assemblies may be categorized as standard draft, high-quality draft, noncontiguous finishes, and finished. An assembly should be at least a high-quality draft in order to be a good target for annotation (Li et al., 2018).

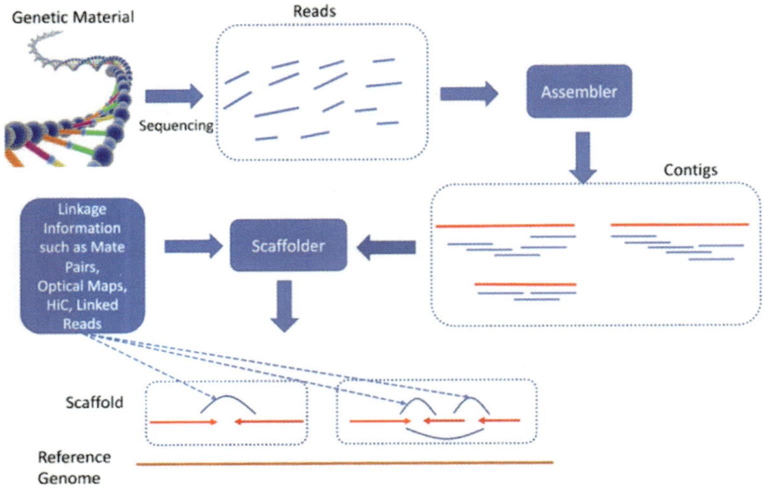

Figure 2.20: *The genome is assembled by sequencing first the genomic DNA producing sequenced fragments called reads. These are then merged by assembler software to generate longer contigs. The contigs are oriented and ordered with respect to each other using a scaffolder software that uses linkage information from various sources such as mate pairs, optical maps, HiC, or linked reads.*

Source: Ghurye and Pop (2019).

2.2.4.1. Assembly challenges

An ideal genome sequencing data have no repetitive sequences, complex structures, and no sequencing errors. In practice, however, there are several factors that may interfere with the assembly such as huge amount of repetitive sequences that are distributed across the whole genome. These repeats may be transposable elements, short tandem repeats, and large segmental duplications (Li et al., 2012). Plant genomes are also characterized with having long terminal repeats made up of identical sequences with hundreds or thousands of copies. This fall into two categories, the Copia elements that are 4–8 kbp long and the Gypsy elements with 8–16 kbp (Li et al., 2018). Moreover, the target DNA is not always extracted from a haploid genome or a homozygous diploid genome but from heterozygous diploid genomes. This becomes even more complicated among plants with polyploid genomes (Li et al., 2012).

The sequencing errors and limited read lengths inherent in the commonly available sequencing technologies add to the difficulty of genome assembly. These challenges are being addressed by the latest sequencing technologies that provide pair-end sequencing, wherein both ends of a fragment are sequenced to improve sequencing accuracy (Li et al., 2012).

2.2.4.2. Data filtering

The sequencing errors in the raw reads from various sequencing platforms can be reduced by filtering and correcting them. Most of these platforms generate a quality value for each base in the reads. A low-quality value indicates a high sequencing error, and this can be used as a basis to filter out certain sequence reads. Nevertheless, some sequencing errors may manifest high quality value which prevents them from being removed from assembly. However, if the genome fragments have been sequenced several times, the correct base is likely to appear with higher frequency than the erroneous ones. This process is called pre-assembly error correction.

OLC assembly software filters low-quality reads but oftentimes skip the pre-assembly error correction. However, this does not usually cause a serious problem during sequencing since the initial step of pair wise alignment to find overlaps allows some mismatches. Also, the multiple sequencing alignments performed to find the consensus sequence errors.

On the other hand, DBG assembly software performs both low-quality filtering and pre-assembly error correction. This is necessary because reads with several sequencing errors will generate false *k*-mers that are not actually part of the genome sequence. These will also bloat the data size consuming more computer memory as well as complicate the graph and make it difficult to infer the contig correctly (Li et al., 2012).

2.2.4.3. Contig assembly

The contigs inferred from the overlapping DNA sequence is eventually mapped by arranging them in order and orientation (see Figure 2.20). One parameter for evaluating genome completeness and contiguity is by determining the N50 contig value. This is determined by first sorting all the contigs in a decreasing manner in terms of size. For instance, if the assembled contigs are 0.33, 1.2, 0.92, 0.85, 0.5, 0.47, and 0.74 Mbp, these are arranged as 1.2, 0.92, 0.85, 0.74, 0.5, 0.47, and 0.33 Mbp. Then these are added to get the total size which, in this case, is 5.01 Mbp. The arranged contigs are again added until the total reaches at least 50% of the total size (i.e., 2.505). The sum of 1.2, 0.92, and 0.85 is already 2.97. Finally, the smallest of these contigs, 0.85, is the N50 contig. The larger the N50 value, the better is the assembly quality (Choudhuri, 2014). An earlier version of the maize genome had a N50 contig value of 19 kbp but the more recent one has 1180 kbp indicating a significant improvement in the quality of the genome assembly (Li et al., 2018).

These contigs are then built into a continuous sequence by using the assembly software. Repetitive sequences are problematic since these have many overlaps with various reads. As a result, the computational time needed for pair-wise alignment is significantly increased in OLC, but not in DBG. This is solved by masking the repeat patterns first when finding the overlaps then these are recovered after contig construction (Li et al., 2012). Finished quality standards or the Bermuda standards consider a contig to be a finished sequence if it has one error per 10,000 bases (Li et al., 2018).

2.2.4.4. Scaffolding

A scaffold is made up of contigs and gaps or links (see Figure 2.20). There are gaps or linkage when one sequenced end of a fragment overlaps with one contig and the other end overlaps with another contig (MycoCosm, 2019) (see Figure 2.21). The scaffold provides information about the long-range structure of the genome without specifying the actual DNA sequence of the gaps between contigs (Ghurye and Pop, 2019). From the length of the fragment, the number of bases between contigs can be estimated (MycoCosm, 2019). This can be several hundreds of base pairs to millions of base pairs depending on the sequencing platform used (Ghurye and Pop, 2019). Repeat contigs can have many links with other contigs making it difficult to infer the relationship between contigs. These repeats can be masked first in order to use only unique contigs to construct scaffolds and then recover the repeat contigs in the end (Li et al., 2012).

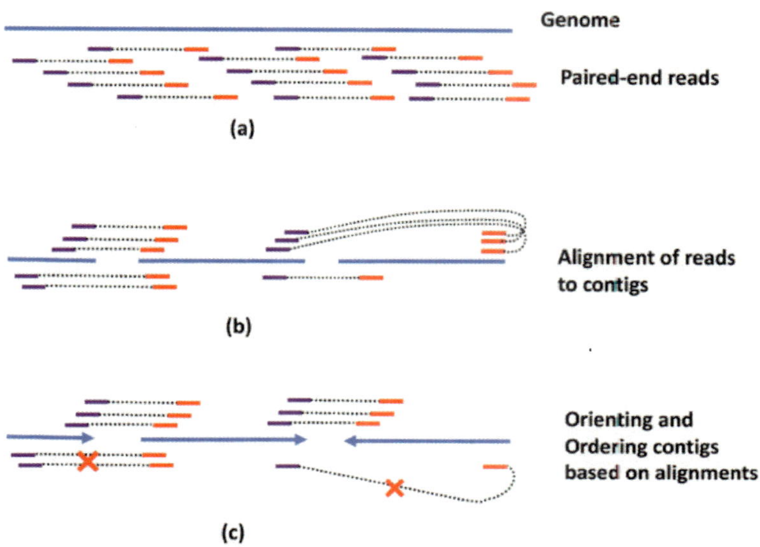

Figure 2.21: *Linkage information is necessary for scaffolding. (a) Paired-end reads are sequenced and then aligned to contigs. (b) Reads with their ends aligned to different contigs provide the linkage information for scaffold assembly. (c) This linkage information is used to orient and order the contigs into the scaffolds.*

Source: Ghurye and Pop (2019).

A chromosome may be represented by one or several scaffolds (MycoCosm, 2019). There are three types of scaffolds: placed scaffold that have been placed within a chromosome, unlocalized scaffolds whose chromosome placement is known but the position or orientation is not known, and unplaced scaffolds which have not been placed in any chromosome (Ensembl Asia, 2019).

It is essential to obtain long-range linkage information to orient and order contigs into scaffolds and eventually infers the chromosome structure (see Figure 2.21). There are certain methods that may be used to obtain this information, which are more efficient and less-labor intensive as opposed to the traditional genetic mapping or physical mapping. One is by preparing DNA library by mate-pair method. In this method, DNA fragments that are more than 1 kbp long are self-ligated to bring the opposite ends together. The resulting circular DNAs are cut and the short fragments containing the joined junction are selected and then sequenced to determine the sequence of the distant parts of the genome (Li and Harkess, 2018).

A useful method for scaffolding eukaryotic DNA which is wrapped around histones is the Hi-C method. Histone packages DNA into a three-dimensional chromatin where two distant regions of the chromatin may be packed close to each other. In the Hi-C method, the DNA is cross-linked to the histones *in vivo* to maintain the chromatin conformation. The DNA is digested allowing physically distant but spatially close fragments to be ligated. Hi-C links fragments within a distance of more than hundreds of kbp. Sequencing of the ligation products follows, and the Hi-C scaffolder uses a statistical model to assemble contigs into individual chromosomes. This method was utilized to improve certain plant genomes such as those of lettuce, barley, and quinoa (Li and Harkess, 2018; Li et al., 2018).

Bio Nano optical mapping method uses restriction endonucleases to create nicks at specific recognition sites in a very long DNA molecule with 150 kbp or more. Fluorescently-labeled nucleotides are incorporated at the nicks. The intervals between the fluorescent labels are visualized allowing the construction of a genome-wide restriction map. This map is used with an *in silico* digestion of the genome assembly to generate linkage information for scaffolding. Several restriction enzymes may be used to produce separate but complementary optical maps. This was used to scaffold the genomes of maize and asparagus (Li and Harkess, 2018). Long reads in the maize reference genome generated 2,958 contigs which were reduced to 625 scaffolds with the use of an optical map. Optical mapping also revealed structural variations in the genome of two inbred maize lines, the W22 and Kill. It showed the prevalence of deletions in low gene density regions and in maize lineage-specific genes (Li et al., 2018).

Another method is 10x genomics. This platform partitions input genome DNA into numerous droplets with a dozen or so DNA molecules per droplet. Within each droplet, library preparation is performed, and a unique barcode is assigned to it. Thus, sequence reads with the same barcode can only come from the few DNA molecules in each droplet allowing the inference of linked reads. One distinct advantage of this method is that homologous DNA molecules are unlikely to be placed in the same droplet so that each haplotype can have its own barcode and can be distinguished (Li and Harkess, 2018).

2.2.4.5. Filling the Gaps

The gaps in the genome need to be closed. Gaps are regions that are difficult to assemble either due to repeats, which form repeat gaps, or due to uncoverage of sequencing, the so-called Lander-Waterman gaps (Li et al., 2012; Walve et al., 2017). Either of these two may form out-gaps, which are found between scaffolds and are often large and cannot be crossed by available pair-end reads, or in-gaps, which are inside scaffolds and are often small and thus, can be crossed by available pair-end reads. Theoretically, repeat gaps can be filled in by retrieving repeat ends and contigs which were not assembled in scaffolds. On the other hand, for Lander-Waterman gaps, the missing reads need to be generated by additional sequencing of the fragments localized in the gap regions. These are often produced by PCR amplification. When the necessary reads and contigs for a gap are obtained, OLC or DBG algorithms are used to close the gap. Since the sequences within gaps are usually repeats that cause aligning problems, gap sequences often have low quality (Li et al., 2012).

Gap-filling modules are included in many genome assembly software. There are also stand-alone gap filling computer programs. The methods employed by these tools vary but all of them perform local genome assembly with a set of reads from the genome (Walve et al., 2017).

2.3. INTERPRETING GENOME DATA

After coming up with the genome sequence, the next step is annotation, a process where the genes, repetitive elements, and the regulatory, non-coding regions in the genome are determined (Li et al., 2018). Genomes may be annotated by expressed sequence tags (ESTs), which are mRNA sequences derived from cDNA libraries (Parkinson and Blaxter, 2009), as well as from protein sequences from related species and from gene finding by computational means (Li et al., 2018). There are computer programs, such as Benchmarking Universal Single-Copy Orthologs (BUSCO), that may be used to identify genes that are typically present in single copy across major lineages. Some of these annotating tools are specific for plant genomes like MAKER-P (Li and Harkess, 2018).

There are three categories of genome annotation. Nucleotide-level annotation attempts to locate the genes and repetitive elements. Then there is protein-level annotation that attempts to determine the possible functions of genes. Finally, process-level annotation determines the pathways and processes in which different genes interact. Sequencing or assembly errors may compromise the accuracy of all three types of annotations possibly resulting in false or pseudo genes or interfering with the inference of true gene function (de Sá et al., 2018). After genome annotation, the genome sequence is usually deposited in GenBank, a database of annotated DNA sequences that are publicly available. This includes a record which describes each gene, its RNA or protein product, and the function of the gene product (see Figure 2.22). The record may also include a summary of the evidence for the assigned function (Koonin and Galperin, 2003).

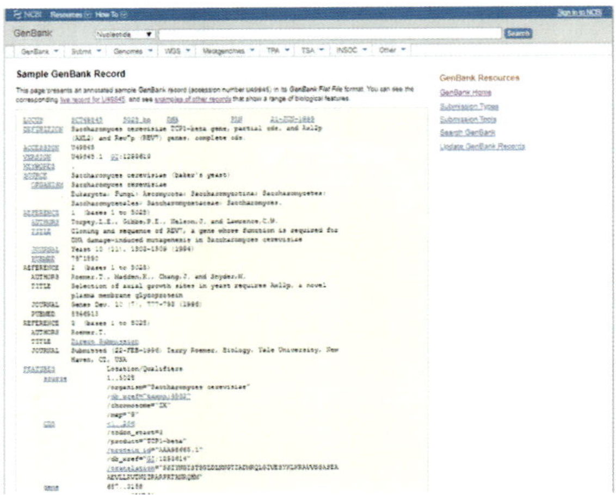

Figure 2.22: *A sample record of a gene that can be accessed in GenBank.*

Source: National Center for Biotechnology Information, U.S. National Library of Medicine, Public Domain.

2.4. *ARABIDOPSIS THALIANA*: MODEL FOR PLANT GENOMICS

Arabidopsis thaliana is a small flowering plant which belongs to the mustard family (Cruciferae) (see opening image). It is widely considered as a model for studying plant biology. Its rapid growth cycle, small size, and large seed output make it ideal for genetics experiments. In addition, its small genome size and low repetitive content also make it a good candidate for the first genomic sequencing project in plants (Goodman et al., 1995).

2.4.1. Early Efforts of Studying *Arabidopsis thaliana* Genome

The initial efforts of mapping the genes of *A. thaliana* involved classical segregation analysis. This produced genetic maps that included approximately 90 genes. These eventually expanded to include additional markers generating a map with more than 280 markers. These markers are significant in conducting genome research in this plant (Goodman et al., 1995).

Several methods were employed to estimate the genome size of *A. thaliana* such as, reassociation kinetics, flow cytometry, chromosome volume, and physical mapping of cosmids. The estimated genome size ranged from 50 to 150 Mbp (Goodman et al., 1995).

2.4.2. Sequencing the *Arabidopsis thaliana* Genome

The genome of *A. thaliana* has 125 Mbp organized into five chromosomes. It contains 25,498 genes that code for 11,000 protein families. Genome sequencing for this plant started in 1996 and was finished in 2000. Three groups are involved in the genome sequencing project, namely European Union Chromosome 4 and 5 Sequencing Consortium, European Chromosome 3 Sequencing Consortium, and The Cold Spring Harbor and Washington University Genome Sequencing Center Consortium (AGI, 2000).

DNA for sequencing was sourced from large-insert, phage (P1), and transformation-competent artificial chromosome (TAC) libraries derived from DNA of accession Columbia-0 (Col-0). Early stages of genome sequencing utilized cosmid clones. Sanger sequencing method was used to complete the genome. The resulting sequences were integrated with the genetic map in order to assemble sets of contigs. The sequences from the 1,569 BAC, TAC, cosmid, and P1 clones were assembled into 10 contigs representing the chromosome arms and the centromere. An additional 22 PCR products were amplified from the genomic DNA and then sequenced to link the regions not covered by the cloned DNA. Certain yeast artificial chromosome (YAC) and phage clones as well as inverse polymerase chain reaction (IPCR) products were derived from genomic DNA and then used to obtain the telomere sequence (AGI, 2000).

The finished genome is considered the gold standard for plant genomes due to its high quality. Still, there are gaps in the 125 Mbp finished genome sequence as it had been experimentally estimated that *A. thaliana* genome is actually 157 Mbp. Expectedly, most of these gaps are the highly repetitive sequences that are the usual problems in sequencing projects, and these represent regions in the centromere and ribosomal DNA (rDNA). However, the gaps also include some euchromatin (Hamilton and Buell, 2012).

There are 25,498 genes that were predicted to be present in the *A. thaliana* genome. Parameters based on known *Arabidopsis* gene structures as well as similarities to known protein and ESTs were used to refine gene models. The three sequencing groups were consistent in their prediction of 83% of the genes. Among these, 93% of the ESTs matched the proposed genes while less than 1% of ESTs matched the predicted non-coding regions. These indicate that most of the potential genes were identified. In addition, 589 cytoplasmic tRNAs, 27 organelle-derived tRNAs, and 13 pseudogenes were identified in the genome. All of the 46 tRNA families required to carry all 61 codons were identified (AGI, 2000).

The functions of 69% of the genes were determined by the sequence similarity of their predicted protein product to proteins of known function in all organisms, such as those involved in metabolism, transcription, transport, and homeostasis. Only 9% of the genes have been characterized experimentally. About 30% of the predicted protein products could not be assigned to any functional category. Comparison of these genes and predicted protein products to those of other organisms indicate that there are genes of *A. thaliana* that are related to other eukaryotic genomes as well as bacterial genomes. Nevertheless, analysis of the *A. thaliana* genome shows that while many protein families are shared between all eukaryotes, plants contain about 150 unique protein families (AGI, 2000).

The *A. thaliana* genome is significantly redundant due to segmental duplications and tandem arrays. These segmental duplications are responsible for 6,303 gene duplications. Specifically, there are 24 large duplicated segments of 100 kbp or longer in the genome. These comprise 58% of the genome. Additionally, there are 1,528 tandem arrays in the genome which contain 4,140 genes. In fact, 17% the *A. thaliana* genes are arranged in tandem arrays. These arrays may have up to 23 adjacent members (AGI, 2000).

Another genome assembly using Illumina sequencing platform was produced using *A. thaliana* accession Niderzenz-1 (Nd-1_v1). This genome was further improved by employing SMRT sequencing data and this was designated as Nd-1_v2C. This version showed 75-fold greater contiguity than Nd-1_v1, with an average contig length of 1.8 Mbp. The completeness of this genome assembly was confirmed by the occurrence of telomeric repeat sequences, as well as the detection of 98.2% of the genes (one more than that detected in the Col-0 genome) via the software BUSCO (Pucker et al., 2019). BUSCO quantifies completeness of genomic data by evaluating expected gene content (Waterhouse et al., 2018). Despite significantly improved contiguity in this genome assembly, centromeric regions with an estimated size of 5 Mbp each were still not assembled. This may be because the error rate of long reads is still too high and so highly repetitive regions such as the centromeres and the nucleolus organizing regions (NORs) cannot be completely resolved (Pucker et al., 2019).

The long sequencing reads can resolve structural variants among different accessions of *A. thaliana* such as larger insertions and deletions that are often missed by short read mappings. Based on the Nd-1_v2C genome sequence assembly, gene duplications and genome rearrangements between Col-0 and Nd-1 were identified. These rearranged segments range in size from a few kbp to 1 Mbp. The 1 Mbp inversion was observed on chromosome 4 (Pucker et al., 2019).

REFERENCES

1. Abizar, *Creative Commons Attribution-Share Alike 3.0 Unported License*. https://en.wikipedia.org/wiki/Sanger_sequencing#/media/File:Radioactive_Fluorescent_Seq.jpg; https://commons.wikimedia.org/wiki/File:Radioactive_Fluorescent_Seq.jpg (Accessed on 18 December 2019).
2. Aluquette. *Creative Commons Attribution-Share Alike 3.0 Unported License*. https://commons.wikimedia.org/wiki/File:Genomic_Library_Construction.png (Accessed on 18 December 2019).
3. Applied Biosystems, (2009). *DNA Sequencing by Capillary Electrophoresis* (2nd edn.). https://tools.thermofisher.com/content/sfs/manuals/cms_041003.pdf?icid=lsg_gsd_CP1236_PJT8083_blog_seqitout-howsangerworks (Accessed on 18 December 2019).
4. Arrow: SimpleIcon. *Creative Commons Attribution 3.0 Unported License*. http://www.flaticon.com/packs/simpleicon-interface (Accessed on 18 December 2019).
5. Bite Size Bio, (2017). *An Introduction to Nanopore Sequencing*. https://bitesizebio.com/36592/introduction-nanopore-sequencing/ (Accessed on 18 December 2019).
6. Choudhuri, S., (2014). *Bioinformatics for Beginners: Genes, Genomes, Molecular Evolution, Databases, and Analytical Tools*. Academic Press: London.
7. Cullis, C. A., (2004). *Plant Genomics and Proteomics*. John Wiley and Sons: New Jersey.
8. David, T. (2011) *Creative Commons Attribution-Share Alike 3.0 Unported License*. https://commons.wikimedia.org/wiki/File:DNTP_nucleotide_incorporation_reaction.svg (Accessed on 18 December 2019).
9. De Sá, P. H. C. G., Guimarães, L. C., Das Graças, D. A., De Oliveira, V. A. A., Barh, D., Azevedo, V., da Costa da Silva, A. L., & Ramos, R. T. J., (2017). Next-generation sequencing and data analysis: Strategies, tools, pipelines and protocols. In: Barh, D., & Azevedo,V., (eds.), *Omics Technologies and Bioengineering: Towards Improving Quality of Life* (pp. 191–207). Academic Press: Massachusetts.
10. Ensembl Asia, (2019). *Chromosomes, Scaffolds and Contigs*. https://asia.ensembl.org/info/genome/genebuild/chromosomes_scaffolds_contigs.html (Accessed on 18 December 2019).

11. Estevezj. *Creative Commons Attribution-Share Alike 3.0 Unported License,* https://en.wikipedia.org/wiki/Sanger_sequencing#/media/File:Sanger-sequencing.svg (Accessed on 18 December 2019).
12. European Bioinformatics Institute, (2019a). *Train Online: 454 Sequencing.* https://www.ebi.ac.uk/training/online/course/ebi-next-generation-sequencing-practical-course/what-next-generation-dna-sequencing/454-seque (Accessed on 18 December 2019).
13. European Bioinformatics Institute, (2019b). *Train Online: Ion Torrent: Proton/PGM Sequencing.* https://www.ebi.ac.uk/training/online/course/ebi-next-generation-sequencing-practical-course/what-next-generation-dna-sequencing/ion-torre (Accessed on 18 December 2019).
14. Farrar, S., (2019). *News Medical Life Sciences: High-throughput DNA Sequencing Techniques.* https://www.news-medical.net/life-sciences/High-throughput-DNA-Sequencing-Techniques.aspx (Accessed on 18 December 2019).
15. Fichot, E. B., & Norman. R. S., (2013). Microbial phylogenetic profiling with the pacific biosciences sequencing platform. Creative commons attribution 2.0 generic license. *Microbiome, 1*(10). doi: 10.1186/2049-2618-1-10.
16. Ghurke, J., & Pop, M., (2019). Modern technologies and algorithms for scaffolding assembled genomes. *PLoS Computational Biology, 15*(6), e1006994. https://doi.org/10.1371/journal.pcbi.1006994 (Accessed on 18 December 2019).
17. Ghurye, J., & Pop, M., (2019). Modern technologies and algorithms for scaffolding assembled genomes. Creative commons attribution 4.0 international license. *PLoS Computational Biology 15*(6), e1006994. https://doi.org/10.1371/journal.pcbi.1006994 (Accessed on 18 December 2019).
18. Gladman, S., (2019). De novo genome assembly for illumina data. *Melbourne Bioinformatics.* https://www.melbournebioinformatics.org.au/tutorials/tutorials/assembly/assembly-protocol/ (Accessed on 18 December 2019).
19. Goodman, H. M., Eckers, J. R., & Dean, C., (1995). The genome of *Arabidopsis thaliana. Proceedings of National Academy of Sciences U.S.A, 92,* 10831–10835.
20. Grehl, C., Kuhlmann, M., Becker, C., Glaser, B., & Grosse, I., (2018).

How to design a whole-genome bisulfite sequencing experiment. Creative commons attribution 4.0 international license. *Epigenomes, 2*(4), 21. https://doi.org/10.3390/epigenomes2040021 (Accessed on 18 December 2019).

21. Griffiths, A. J. F., Gelbart, W. M., Miller, J. H., & Lewontin, R. C., (1999). *Modern Genetic Analysis*. W. H. Freeman: New York.
22. Hamilton, J. P., & Buell, C. R., (2012). Advances in plant genome sequencing. *The Plant Journal*, *70*(1), 177–190. https://doi.org/10.1111/j.1365-313X.2012.04894.x (Accessed on 18 December 2019).
23. Head, S. R., Komori, H. K., LaMere, S. A., Whisenant, T., Van Nieuwerburgh, F., Salomon, D. R., & Ordoukhanian, P., (2014). Library construction for next-generation sequencing: Overviews and challenges. *BioTechniques*, *56*(2), 61–77. doi: 10.2144/000114133.
24. Heslop-Harrison, J. S., (2003). Plant genomes. Thomas, B., (ed.), *Encyclopedia of Applied Plant Sciences* (pp. 130–133). Academic Press: Massachusetts.
25. Hodgkin, J., (2001). Genome size. Brenner, S., & Miller, J. H., (eds.), *Encyclopedia of Genetics* (p. 865). Academic Press: Massachusetts.
26. Inglis, P. W., Pappas, M. D. C. R., Resende, L. V., & Grattapaglia, D., (2018). Fast and inexpensive protocols for consistent extraction of high quality DNA and RNA from challenging plant and fungal samples for high-throughput SNP genotyping and sequencing applications. Creative commons attribution 4.0 international license. *PLoS One, 13*(10), e0206085. https://doi.org/10.1371/journal.pone.0206085 (Accessed on 18 December 2019).
27. Jacopo, P. (2014) *Density Design Research Lab, Creative Commons Attribution-Share Alike 4.0 International License*. https://commons.wikimedia.org/wiki/File:How_Pyrosequencing_Works.svg (Accessed on 18 December 2019).
28. John, S. *Creative Commons Attribution-Share Alike 3.0 Unported License*. https://en.wikipedia.org/wiki/File:Sequencing.jpg (Accessed on 18 December 2019).
29. KhanAcademy, (2016). *Biotechnology: DNA Sequencing*. https://www.khanacademy.org/science/high-school-biology/hs-molecular-genetics/hs-biotechnology/a/dna-sequencing (Accessed on 18 December 2019).
30. Kim, H., Jebrail, M. J., Sinha, A., Bent, Z. W., Solberg, O. D., Williams, K. P., et al., (2013). A microfluidic cDNA library preparation platform

for next-generation sequencing. Creative commons attribution 4.0 international license. *PLoS One, 8*(7), e68988. doi: 10.1371/journal.pone.0068988.

31. Knief, C., (2014). Analysis of plant microbe interactions in the era of next generation sequencing technologies. Creative commons attribution 3.0 unported license. *Frontiers in Plant Science* 5, 216. https://doi.org/10.3389/fpls.2014.00216 (Accessed on 18 December 2019).
32. Koonin, E. V., & Galperin, M. Y., (2003). *Sequence-Evolution-Function: Computational Approaches in Comparative Genomics*. Kluwer Academic: Boston.
33. Li, C., Lin, F., An, D., Wang, W., & Huang, R., (2018). Genome sequencing and assembly by long reads in plants. *Genes (Basel)*, *9*(1), 6. doi: 10.3390/genes9010006.
34. Li, F. W., & Harkess, A., (2018). A guide to sequence your favorite plant genomes. Creative commons attribution 4.0 international license. *Applications in Plant Sciences, 6*(3), e1030. https://doi.org/10.1002/aps3.1030 (Accessed on 18 December 2019).
35. Li, Z., Chen, Y., Mu, D., Yuan, J., Shi, Y., Zhang, H., Gan, J., Li, N., Hu, X., Liu, B., Yang, B., & Fan, W., (2012). Comparison of the two major classes of assembly algorithms: Overlap-layout-consensus and de-brujin-graph. *Briefings in Functional Genomics*, *11*(1), 25–37. https://doi.org/10.1093/bfgp/elr035 (Accessed on 18 December 2019).
36. Lu, Y., Shen, Y., Warren, W., & Walter, R., (2016). *Next Generation Sequencing in Aquatic Models*. Next generation sequencing-advances, Applications, and Challenges. Creative Commons Attribution 3.0 Unported License. https://www.intechopen.com/books/next-generation-sequencing-advances-applications-and-challenges/next-generation-sequencing-in-aquatic-models (Accessed on 18 December 2019).
37. Mandrioli, M., & Manicardi, G. C., (2012). Unlocking holocentric chromosomes: New perspectives from comparative and functional genomics? Creative commons attribution 2.5 generic license. *Current Genomics, 13*(5), 343–349. doi: 10.2174/138920212801619250.
38. Marie-Lan Nguyen. *Creative Commons Attribution 2.5 Generic License*, https://commons.wikimedia.org/wiki/File:Arabidopsis_thaliana_JdP_2013-04-28.jpg (Accessed on 18 December 2019).
39. Martin, C., Viruel, M. A., Lora, J., & Hormaza, J. I., (2019). Polyploidy

in fruit tree crops of the genus *Annona* (Annonaceae). *Creative Commons Attribution 4.0 International License, Frontiers in Plant Science, 10*, 99. https://doi.org/10.3389/fpls.2019.00099 (Accessed on 18 December 2019).

40. Merck, (2019). *Sanger Sequencing Steps and Methods*. https://www.sigmaaldrich.com/technical-documents/articles/biology/sanger-sequencing.html (Accessed on 18 December 2019).
41. Michael, T. P., (2014). Plant genome size variation: Bloating and purging DNA. *Briefings in Functional Genomics, 13*(4), 308–317. https://doi.org/10.1093/bfgp/elu005 (Accessed on 18 December 2019).
42. MycoCosm., (2019). *What is a Scaffold?* https://mycocosm.jgi.doe.gov/help/scaffolds.jsf (Accessed on 18 December 2019).
43. *National Center for Biotechnology Information*. U.S. National Library of Medicine, Public Domain. https://www.ncbi.nlm.nih.gov/genbank/samplerecord/ (Accessed on 18 December 2019).
44. National Institute of Standards and Technology, (2012). *NIST, Columbia Engineering Collaborate on Inexpensive DNA Sequencing Method*. Public Domain. https://www.nist.gov/news-events/news/2012/10/nist-columbia-engineering-collaborate-inexpensive-dna-sequencing-method (Accessed on 18 December 2019).
45. National Science Foundation, (2000). *First-Ever Complete Plant Genome Sequence is Announced: International Team Reveals DNA Secrets of Arabidopsis thaliana*. https://www.nsf.gov/od/lpa/news/press/00/pr0094.htm (Accessed on 18 December 2019).
46. OpenStax CNX. *Creative Commons Attribution 4.0 License*. https://cnx.org/contents/GFy_h8cu@10.7:5l844Z38@7/Whole-Genome-Sequencing (Accessed on 18 December 2019).
47. Oxford Nanopore Technologies, (2018). *How it Works*. https://nanoporetech.com/how-it-works (Accessed on 18 December 2019).
48. Park, S. T., & Kim, J., (2016). Trends in next-generation sequencing and a new era for whole genome sequencing. *International Neurourology Journal, 20*(2), S76–83, https://doi.org/10.5213/inj.1632742.371 (Accessed on 18 December 2019).
49. Parkinson, J., & Blaxter, M., (2009). Expressed sequence tags: An overview. In: Parkinson, J., (ed.), *Expressed Sequence Tags (ESTs). Methods in Molecular Biology (Methods and Protocols)* (Vol. 533, pp. 1–12). Humana Press: New Jersey.

50. Pearson, P. L., (2001). Triploidy. In: Brenner, S., & Miller, J. H., (eds.), *Encyclopedia of Genetics* (pp. 2055–2056). Academic Press: Massachusetts.

51. Peterson, D. G., & Arick, II, M., (2018). Sequencing plant genomes. In: Cánovas, F., Lüttge, U., Matyssek, R., & Pretzsch, H., (eds.), *Progress in Botany* (Vol. 80, pp. 109–193). Springer: Cham, Switzerland.

52. Pucker, B., Holtgräwe, D., Stadermann, K. B., Frey, K., Huettel, B., Reinhardt, R., & Weisshaar, B., (2019). A chromosome-level sequence assembly reveals the structure of the *Arabidopsis thaliana* Nd-1 genome and its gene set. *PLoS One*, *14*(5), e0216233. https://doi.org/10.1371/journal.pone.0216233 (Accessed on 18 December 2019).

53. Radford, A. D., Chapman, D., Dixon, L. K., Chantrey, J., Darby, A. C., & Hall, N., (2012). Application of next-generation sequencing technologies in virology. Creative commons attribution 2.5 generic license. *Journal of General Virology, 93*(9), 1853–1868. doi: 10.1099/vir.0.043182-0.

54. Raghavendra, P., & Pullaiah, T., (2018). *Advances in Cell and Molecular Diagnostics*. Academic Press: London, U. K.

55. Rhoads, A., & Au, K. F., (2015). PacBio sequencing and its applications. *Genomics, Proteomics and Bioinformatics*, *13*(5), 278–289. https://doi.org/10.1016/j.gpb.2015.08.002 (Accessed on 18 December 2019).

56. Ribeiro, T., Barrela, R. M., Bergès, H., Marques, C., Loureiro, J., Morais-Cecílio, L., & Paiva, J. A. P., (2016). Advancing *Eucalyptus* genomics: Cytogenomics reveals conservation of *Eucalyptus* genomes. *Creative Commons Attribution 4.0 International License*, *Frontiers in Plant Science, 7*, 510. https://doi.org/10.3389/fpls.2016.00510 (Accessed on 18 December 2019).

57. Robin, J. D., Ludlow, A. T., LaRanger, R., Wright, W. E., & Shay, J. W., (2016). Comparison of DNA quantification methods for next generation sequencing. *Scientific Reports*, *6*(24067).

58. Roche, (2019). *Nanopore Sequencing*. https://sequencing.roche.com/en/technology-research/technology/nanopore-sequencing.html (Accessed on 18 December 2019).

59. Schoales, J., (2015). *How Does Sanger Sequencing Work?* https://www.thermofisher.com/blog/behindthebench/how-does-sanger-sequencing-work/ (Accessed on 18 December 2019).

60. Scitable by Nature Education, (2014). *Definition: DNA Sequencing*.

https://www.nature.com/scitable/definition/dna-sequencing-205 (Accessed on 18 December 2019).

61. Snipacademy, (2019). *Semiconductor Sequencing.* https://binf.snipcademy.com/lessons/ngs-techniques/ion-torrent-ph-semiconductor (Accessed on 18 December 2019).

62. The Arabidopsis Genome Initiative, (2000). Analysis of the genome sequence of the flowering plant *Arabidopsis thaliana. Nature, 408*, 796–815.

63. Viv Rolfe. *Creative Commons Attribution-Share Alike 4.0 International License.* https://commons.wikimedia.org/wiki/File:Spectrophotometer_Model_1.JPG (Accessed on 18 December 2019).

64. Walve, R., Salmela, L., & Mäkinen, V., (2017). Variant genotyping with gap filling. *PLoS One*, *12*(9), e0184608. https://doi.org/10.1371/journal.pone.0184608 (Accessed on 18 December 2019).

65. Waterhouse, R. M., Seppey, M., Simão, F. A., Manni, M., Ioannidis, P., Klioutchnikov, G., Kriventseva, E. V., & Zdobnov, E. M., (2018). BUSCO applications from quality assessments to gene prediction and phylogenomics. *Molecular Biology and Evolution*, *35*(3), 543–548. https://doi.org/10.1093/molbev/msx319 (Accessed on 18 December 2019).

66. Zhou, X., & Li, Y., (2015). *Atlas of Oral Microbiology: From Healthy Microflora to Disease* (pp. 15–40). Academic Press: London, U. K.

67. Zimin, A. V., Marçais, G., Puiu, D., Roberts, M., Salzberg, S. L., & Yorke, J. A., (2013). The MaSuRCA genome assembler. *Bioinformatics*, *29*(21), 2669–2677. doi: 10.1093/bioinformatics/btt476.

Chapter 3

Genetic Engineering in Plants

CONTENTS

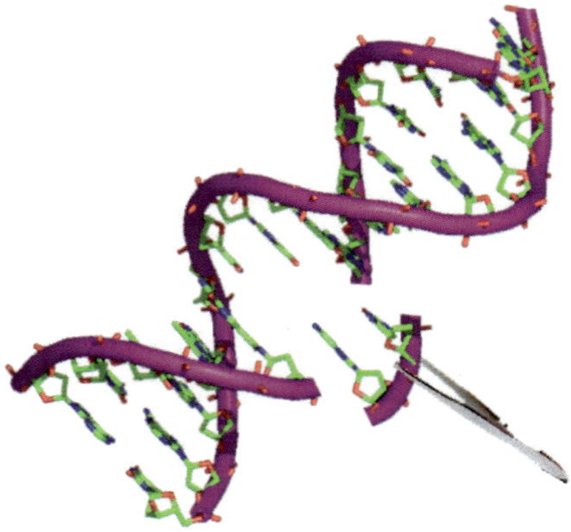

(From Ciencias Españolas and KoS, Creative Commons License).

Animals, plants, and microbes have been modified genetically for centuries now in order to produce desired traits, especially in terms of food production. Such modifications resulted to higher yields, enhanced taste and nutritional value, and increased resistance to biotic and abiotic stresses. The earliest method employed by our ancestors is simple selection where individuals with the most desired traits are chosen for continued reproduction. Another method employed is crossing where two individuals with "superior" traits are made to mate and reproduce a hybrid that possesses the traits from both parents. Advances in methodology and techniques employed were made over the years but the principles of selection and crossing remain to be observed (National Academy of Sciences, 2004).

Genetic engineering is another method of modifying organisms at the DNA level. It involves introducing an exogenous sequence in order to obtain the desired trait or phenotype. This essentially alters the genetic make-up of the organism. This section will discuss the procedure and application of this technology.

3.1. TECHNIQUES OF PLANT GENETIC MANIPULATION

Introducing a gene that does not occur naturally in an organism results to transgenic or genetically-modified organisms (GMO). There are several ways by which a foreign gene is introduced into a plant's genome. These could either be mediated by biological agents such as bacterial or viral vectors or by non-biological means such as various physical or chemical processes. Once the transgene is inside the target cell, it can be expressed transiently or stably if it was integrated into the genome.

3.1.1. *Agrobacterium tumefaciens*-Mediated Transformation

Agrobacterium tumefaciens is a soil bacterium that causes crown gall disease among plants. It contains a tumor-inducing or Ti plasmid (see Figure 3.1). When it infects a susceptible plane host, it transfers a part of its Ti plasmid, called the T-DNA or T-region, into the plant cell (see Figure 3.2). This is about 10 to 30 kbp in size and there may be one or more T-DNA in a Ti plasmid. The T-DNA is then integrated into the plant's chromosome (see Figure 3.3.) so that the host's own cellular machinery will transcribe and translate the bacterium's genes resulting to the production of substances that mediate crown gall development (National Academy of Sciences, 2004; Tzotzos et al., 2009; Gelvin, 2003).

In the early 1980s, strains of *A. tumefaciens* were developed that lacked the disease-causing genes in their Ti-plasmid but retained its ability to attach to the host plant cell and transfer its T-DNA (National Academy of Sciences, 2004). These "disarmed" *A. tumefaciens* plasmids are then made to contain the gene of interest. Specifically, the gene of interest and an antibiotic resistance gene are placed between the left and right borders of the T-DNA (see Figure 3.2 and Figure 3.3). The transformed plasmids are then reintroduced into *A. tumefaciens*. This *A. tumefaciens* is cultivated along with embryogenic suspensions or callus of the desired host plants resulting in the infection of the plant and thus, allowing the modified T-DNA to integrate into the host's genome (Tzotzos et al., 2009).

Figure 3.1: *Agrobacterium tumefaciens is a soil-borne pathogen that causes crown gall disease which is characterized by tumorous growth in plant roots.*

Source: C.M., Creative Commons License.

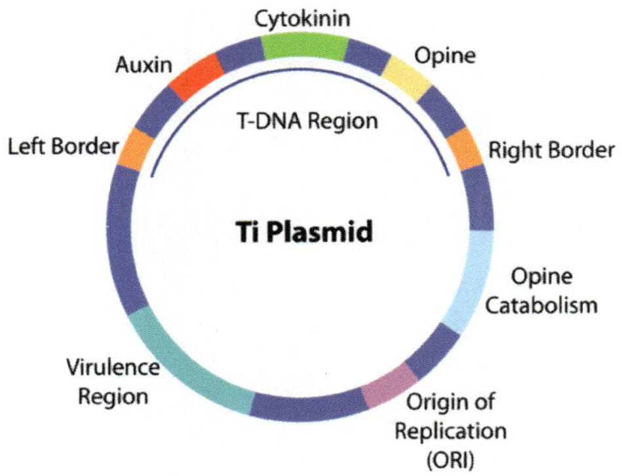

Figure 3.2: *Agrobacterium tumefaciens has tumor-inducing (Ti) plasmid with a T-DNA region which is incorporated into the susceptible host plant's chromosome upon infection.*

Source: Mouagip, Public Domain.

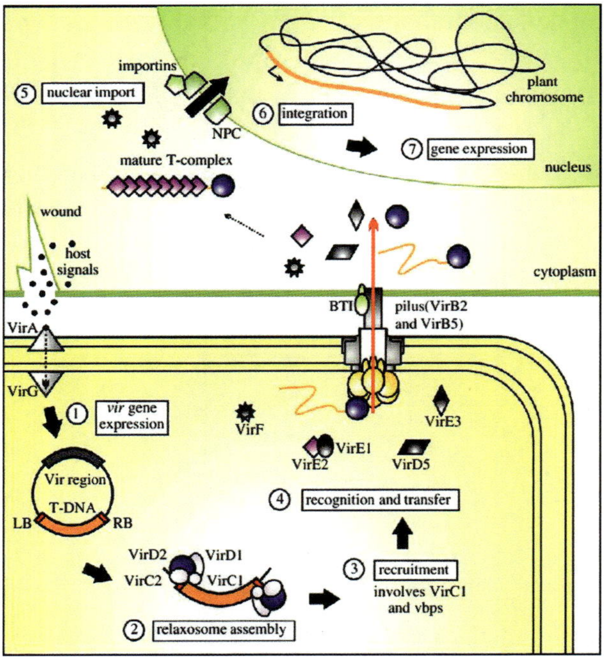

Figure 3.3: *Agrobacterium tumefaciens-mediated transfer of the T-DNA involves several stages. (1) the VirA/VirG system in the bacterium induces the expression of the vir gene in the Ti plasmid. This is in response to signal molecules at wound sites. (2) VirD2 relaxase and relaxsome components excise the*

T-DNA at the left border and right border (LB and RB) sequences. (3) The T-DNA-relaxase complex is recruited to the type IV secretion system (T4SS) which enables the complex to traverse the bacterial envelope and be translocated to the host cell. (4) T4CP VirD4 (yellow) recognizes the T-DNA-relaxase complex and initiates its transfer to the plant cytosol which is followed by (5) its import into the nucleus, (6) integration into the host chromosome, and (7) expression of the transferred gene.

Source: Zechner, Lang, and Schildbach (2012).

The presence of the gene of interest in the target plant is confirmed by growing it in a medium with a certain antibiotic so that only the calli that were successfully infected and integrated with the T-DNA bearing both the gene of interest and gene for resistance to that antibiotic will grow. These eventually develop into plantlets that can be replanted (see Figure 3.4).

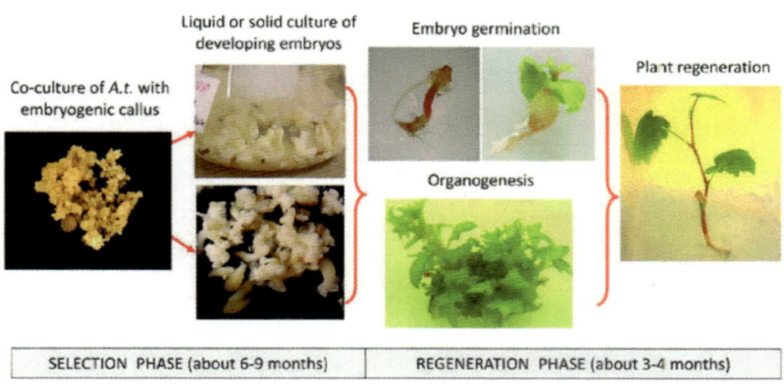

Figure 3.4: *The genetically engineered Agrobacterium tumefaciens is co-cultured with the callus of the target plant in order to allow the bacterium to infect the plant cells and eventually transfer the T-DNA bearing a gene of interest and a gene for antibiotic resistance. The culture medium contains antibiotic so that only the callus with the T-DNA will survive, grow, and regenerate.*

Source: Dalla, Malnoy, and Gribaudo (2017).

A. tumefaciens has been used to genetically transform several dicot species. Most monocots are outside the host range of *A. tumefaciens,* but it had been used in some monocot angiosperms and gymnosperms. It was also used to transform non-plant organisms such as fungi, algae, sea urchin, and even human cells (Hwang et al., 2017).

3.1.1.1. The Ti plasmid

As described previously, the T-DNA is defined by the right and left border sequences (see Figure 3.2). These borders are made up of highly homologous sequence composed of 25 bp. They flank the T-DNA in a directly repeated orientation (Gelvin, 2003).

The virulence region contains at least seven *vir* genes (*vir A, vir B, vir C, vir D, vir E, vir F,* and *vir G*) involved in the process of T-DNA processing and transfer (see Figure 3.2 and Figure 3.3). The Vir A and Vir G proteins make up a sensory-signal regulatory system that activates the expression of the other *vir* genes. Vir A is a transmembrane protein that detects the presence of certain plant phenolic compounds, such as lignin and flavonoid precursors, that are produced upon wounding (Gelvin, 2003; Hwang et al., 2017). In the presence of these compounds and sugar molecules, Vir A autophosphorylates and then transphosphorylates the Vir G protein. Upon its phosphorylation, Vir G increases the transcription rate of the *vir* genes (Gelvin, 2003). Thus, the presence of phenolic compounds is necessary to stimulate virulence gene expression in *A. tumefaciens* (Hwang et al., 2017).

The border sequences are cleaved by the VirD1/VirD2 border-specific endonuclease. This results to a nick on the bottom strand of the T-DNA between the third and fourth bases of the border sequences. Then VirD2 protein is tightly linked to the 5' end at the right border nick of the resulting single-stranded T-DNA molecule. This T-strand complex, not a double-stranded T-DNA molecule, is transferred to the host plant cell (Gelvin, 2003; Hwang et al., 2017).

The VirD4 protein and the 11 Vir B proteins constitute the type IV secretion system (T4SS), which enables the transfer of the T-DNA and Vir proteins from the *A. tumefaciens* to the plant host. VirD4 is an integral membrane protein that may facilitate the interaction of the T-DNA/VirD2 complex with the Vir B proteins. Most of the Vir B proteins form the membrane transporter, with an inner membrane translocase (VirB3-B4-B6-B8-B11) and outer membrane core complex (VirB7-B9-B10). VirB2-B5 proteins make up the extracellular pilus, which may serve as a linker for T-DNA and Vir protein transfer or may function as a hook that brings the *A. tumefaciens* and the host plant into close proximity in order to have a successful infection. The VirD4 protein has ATPase activity that provides the energy for presenting the T-DNA and protein substrates to the Vir B transporter (Gelvin, 2003; Hwang et al., 2017).

VirD2, which forms a complex with the T-DNA to be transferred to the host during infection, may serve to guide the T-strand to and through the T4SS apparatus. It is also proposed that VirE2 facilitates the transfer of the T-strand complex by forming a pore in the plant membrane. Once inside the plant cell, the nuclear localization signal (NLS) sequences contained in VirD2 and VirE2help direct the T-DNA to the plant nucleus. Both VirD2 and VirE2 need to be associated with T-DNA inside the plant cell before the T-DNA can be successfully directed into the plant nucleus. When VirE2 associates with the T-DNA, it can change the conformation of T-DNA from a random-coil to a form that resembles a telephone coil. It can also protect the T-DNA from being digested by nucleases in the plant cytoplasm. This new form may help direct the T-DNA into the nuclear pore. VirD2 and VirE2 may also have a role in the incorporation of the T-DNA into the host chromosome (Gelvin, 2003; Hwang et al., 2017).

After the successful import of the T-strand complex into the plant nucleus, the associated virulence proteins need to be removed to enable efficient T-DNA integration. This is accomplished by the disassembly and targeted proteolysis of the complex, which are mediated by Vir F protein and the host plant's ubiquitin proteasome complex (Hwang et al., 2017).

3.1.1.2. Introducing gene of interest into the T-DNA

There are two strategies of introducing the gene of interest into the T-DNA. One is by indirect cloning where the gene is in *cis* (i.e., in the same strand or plasmid) as the *vir* genes. Another approach is by making T-DNA binary vectors where the gene is cloned into a T-region that is separate from the one containing the *vir* genes (Gelvin, 2003).

Two methods are used to clone foreign DNA into the Ti plasmid. In the first method, a region of DNA containing unique restriction endonuclease sites is cloned into a broad-host-range plasmid. This initial cloning step is performed in *Escherichia coli* where the plasmid vector replicates. Next, the foreign gene of interest and an antibiotic resistance gene (serving as a selection marker gene) are cloned into a unique restriction endonuclease site within the target region of DNA. The plasmid produced is introduced into *A. tumefaciens* either by conjugation or transformation. Successfully transformed *A. tumefaciens* are selected by plating them on a culture medium with certain antibiotics. This step ensures that only bacteria containing the recombinant plasmid, which has an inherent antibiotic resistance gene and the resistance marker gene near the gene of interest, will survive and replicate (see Figure 3.4). At this point, the selected *A. tumefaciens* has two plasmids, its own Ti plasmid and the recombinant plasmid. Then another plasmid belonging to the same incompatibility group as the first plasmid used, with its own antibiotic resistance marker gene, is introduced into the previously selected *A. tumefaciens*. The resulting bacteria are then cultured on medium with two antibiotics in order select for resistance markers next to the gene of interest and in the second plasmid. Since plasmids belonging to the same incompatibility group cannot occur together in the same bacterium, resistance to both antibiotics indicates that either the first plasmid integrated into the Ti plasmid or the first plasmid and the Ti plasmid exchanged DNA by double homologous recombination. Double homologous recombination utilizes the homologous sequences that flank the gene of interest and the resistance marker gene. In this method, the inherent antibiotic resistance gene in the plasmid vector will not be integrated into the Ti plasmid (Gelvin, 2003).

Another method of *cis*-integration of the gene of interest into the Ti plasmid involves an origin of DNA replication found in the colicin E1 plasmid of *E. coli* (ColE1 replicon) (Oxford Reference, 2019). ColE1 replicon, such as pBR322 (pBR322 is one of the widely-used *E. coli* plasmid vectors for cloning), is introduced into the T-DNA of the Ti plasmid resulting to a modified *A. tumefaciens*. On the other hand, the gene of interest is cloned into a separate pBR322-derived plasmid which contains a second antibiotic resistance marker gene. This plasmid is then introduced into the modified *A. tumefaciens*. The resulting strains are then grown on a culture medium that contains the antibiotic corresponding to the second antibiotic resistance marker in the pBR322-derived plasmid. ColE1 replicon cannot function in *A. tumefaciens*, which means that the second antibiotic resistance marker can only be stably expressed if the pBR322-derived plasmid co-integrated into the pBR322 segment of the altered T-DNA (Gelvin, 2003).

The binary-vector system is an alternative strategy that was developed due to the tediousness of the *cis*-integration approach of introducing the gene of interest into the T-DNA. In this approach, the T-DNA and the *vir* genes are separated into two different replicons. The small plasmid containing the T-DNA with the gene of interest is the binary vector while the plasmid with the *vir* genes is the helper or accessory plasmid. The latter generally contains a complete or partial deletion of the T-DNA so that strains with this plasmid cannot induce tumor formation. When these two replicons are in the same *A. tumefaciens* bacterium, the products of the *vir* genes function for T-DNA processing and transfer to the host plant cell (Gelvin, 2003).

An improved ternary vector system was developed using the pVIR accessory plasmids (see Figure 3.5). These are disarmed Ti plasmids which are desirable vectors due to their small size, enhanced stability, improved bacterial selectable marker, and modified *vir* genes for improved T-DNA delivery. The T-DNA is contained in an independent binary vector. This method simplified vector construction and has proven successful in improving the frequency and efficiency of genetic transformation in maize, sorghum, and wheat (Anand et al., 2018; Anand and Jones, 2018).

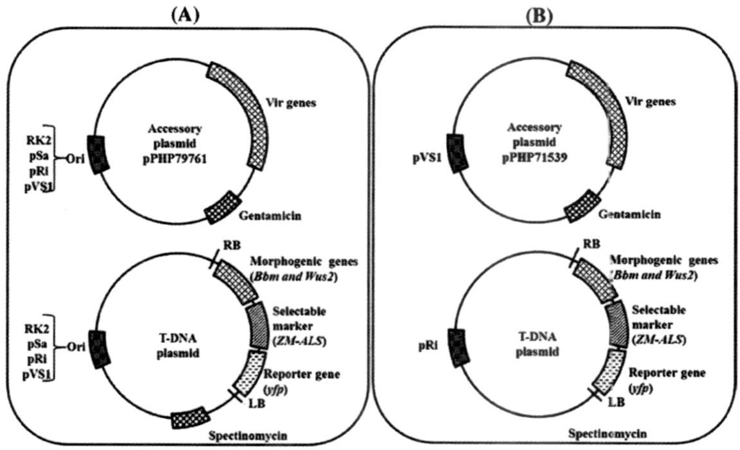

Figure 3.5: *Ternary vector systems used in the rapid transformation of maize include pVIR accessory plasmids (A: pPHP79761 and B: pPHP71539), which contain corrected or additional vir genes to improve the transformation process and are compatible with various types of bacterial origin of replication (Ori). The binary vector contains the selectable antibiotic resistance gene (for the antibiotic spectinomycin) as well as the T-DNA which, apart from the gene of interest, may include morphogenic genes, a reporter gene, and a selectable marker gene. The morphogenic genes Baby boom (Bbm) and Wuschel2 (Wus2) facilitate transformation by allowing the direct formation of somatic embryos without the intermediate callus phase.*

Source: Anand et al. (2018).

3.1.1.2.1. Integration of Unwanted DNA Segments

While the goal is to transfer only T-DNA, which has the gene of interest and certain selection marker genes, it is unavoidable that plasmid backbone sequences are also carried over and integrated into the host plant genome. These unwanted sequences may interfere with the regulation of expression of the gene of interest as well as the endogenous genes in the host plant. It was demonstrated that incorporating toxic "Killer" gene into the binary vector backbone sequences can significantly reduce the percentage of integrated transgenic DNA with such extra sequences (Gelvin, 2003).

3.1.1.2.2. Removal of the Antibiotic Resistance Genes

From the previous discussion, it is apparent that the incorporation of genes for antibiotic resistance into the cloning vectors is integral for screening successfully transformed *A. tumefaciens* strains. However, there are concerns regarding the spread or escape of antibiotic resistance genes in nature. In order to address these, methods of removing these resistance markers have been devised (Gelvin, 2003).

Some of the methods include the use of site-specific recombinase to remove the marker, transposon-based movement of the selection marker from its insertion site in the plant genome to another unlinked site from which it can be segregated in future generations, or the use of multiple T-DNAs which can insert into unlinked sites for subsequent segregations (Gelvin, 2003).

3.1.2. The Use of Viral Vectors

Some of the viruses used as vectors for plant genetic manipulation include DNA viruses like the wheat dwarf virus or cabbage leaf curl virus and RNA viruses such as the tobacco rattle virus, tobacco mosaic virus or potato virus x. These have proven efficient in introducing genetic materials in tobacco, potato, tomato, rice, and wheat (Zaidi and Mansoor, 2017; Darbani et al., 2008). Recombinant viral RNA or viral cDNA, in which an antibiotic resistance gene replaces the viral coat protein gene and which contains the target genetic material, can be inoculated directly into the plants or introduced via agroinoculation or agroinfiltration. Agroinoculation is a process in which the recombinant viral cDNAs are incorporated into the T-DNA of *A. tumefaciens* Ti plasmid and introduced into the plant via the bacterium. When the viral cDNAs are expressed, recombinant viruses are produced which can then infect and integrate the gene of interest into the host plant (see Figure 3.6) (Gelvin, 2005).

Direct inoculation of viral vectors can be done rapidly, and it has shown high transient gene expression. The viral vector can also spread throughout the host plant and so it has the potential to be used in transforming an entire plant (Darbani et al., 2008). Thus, the transformed plant can become a “factory” to produce certain gene products (Gelvin, 2005).

However, vectors containing average length and long transgenes showed low infectivity, like in the case of agroinoculation which infects only a small number of cells. In this case, only a small amount of the recombinant proteins will be produced (Gelvin, 2005). It is possible that introduction of the viral cDNA into the target cell via *A. tumefaciens* directs it to the nucleus which is not usually reached by viruses during their life cycle, since they normally stay in the infected cell’s cytoplasm. In the nucleus, viral RNA is forced to go through pre-mRNA processing which may splice the RNA into a form that can no longer replicate or spread throughout the plant. When possible cryptic splice sites and T-rich intron-like sequences in the viral vector, which are possible targets of plant nuclear RNA processing, are altered by introducing neutral mutations, this can significantly increase viral infectivity (Marillonet et al., 2005).

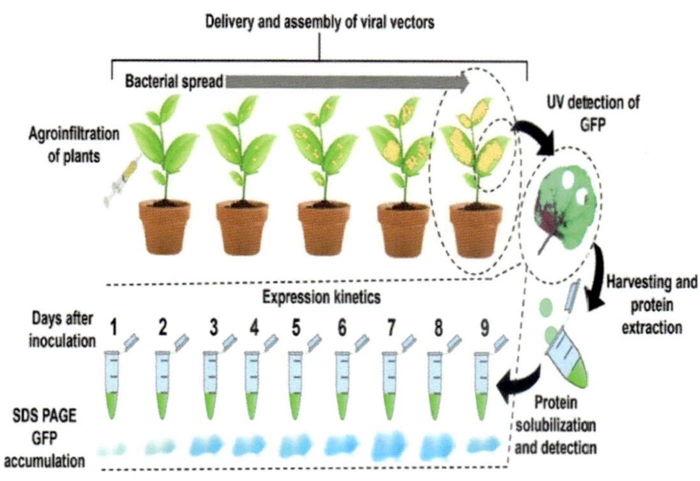

Figure 3.6: *Agroinoculation employs Agrobacterium tumefaciens to deliver viral vectors containing the gene of interest into the target plant. Expression of the viral vectors result to recombinant viruses being produced and infecting the whole plant and integrating the gene of interest into the host plant. As a result, the transformed plant produces the recombinant protein.*

Source: Leite et al. (2019).

Another possible reason for the low infectivity of agroinoculation is that the nuclear RNA processing and export machinery have difficulty recognizing the viral RNA with more than 7 kbp because it normally recognizes plant mRNAs which are normally 1–2 kbp. Introducing up to 14 introns into both the viral vector and the gene of interest made the viral RNA resemble mRNA. These additions along with the previous modifications can increase the infectivity further so that more than 90% of the cells are expressing the transgene (Marillonet et al., 2005).

Other disadvantages of using viral vectors include low stable transformation efficiency, integration of multiple copies of the gene of interest, chromosomal rearrangement, and appearance of disease symptoms (Darbani et al., 2008).

3.1.3. Microprojectile Bombardment

Microprojectile bombardment or the use of a gene gun delivers naked DNA to plant cells by shooting them at high velocity with microscopic metal pellets coated with DNA (see Figure 3.7) (National Academy of Sciences, 2004; Christou, 1992). This is not as efficient as *A. tumefaciens*-mediated transformation, but it proved to be useful in genetically modifying plants that are not susceptible to *Agrobacterium* infection like most monocots such as maize, rice, and other cereal grains as well as in transforming various tissue and cell types (National Academy of Sciences, 2004; Baltes et al., 2017). In addition, this method does not require binary vectors, allows the delivery of multiple genes of interest, and bypasses tissue culture-related difficulties in regenerating transformed plant cells (Baltes et al., 2017; Christou, 1992).

Several ways can be used to accelerate particles into living cells like pneumatic devices; instruments using either mechanical impulse, centripetal, magnetic, or electrostatic forces; spray or vaccination guns or apparatus that accelerates particles by shock wave (Christou, 1992).

The metal particles may be coated with DNA or RNA or they may carry the genetic material from a solution of DNA or RNA surrounding the cell. The DNA or RNA of interest may be released immediately upon entry into the cytoplasm or nucleus or be made available through a slow release mechanism. The metal particles to be utilized should have enough mass so that they have sufficient momentum to penetrate into the target tissue. Additionally, these particles should be chemically inert in order not to have adverse reaction with the DNA or cell components, but they should be able to form transient complexes with the DNA molecule, which can be dissociated once the particles enter the cell. Examples of metal particles suitable for this purpose include gold, tungsten, palladium, rhodium, platinum, and iridium (Christou, 1992).

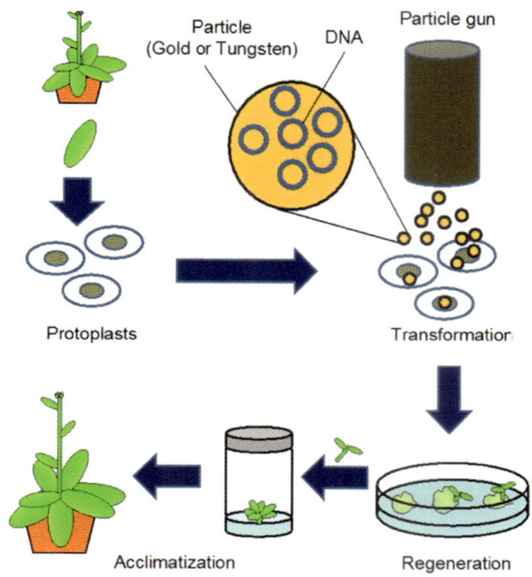

Figure 3.7: *Microprojectile bombardment requires the isolation of leaf protoplasts (these are cells growing in culture with their cell walls removed) where the DNA-coated particles are injected. After successful genetic transformation, the protoplasts are regenerated into whole transgenic plants. These are then acclimatized in a greenhouse.*

Source: Narusaka, Narusaka, Yamasaki, and Iwabuchi (2012).

It is also important to ensure that the appropriate target cells or tissues are competent for transformation and regeneration. In the case of a tissue sample, the depth of penetration needs to be fine-tuned in order to achieve particle delivery to specific cell layers (Christou, 1992). Even by controlling such factors, the integration patterns and the final destination of the injected particles (whether in the cytoplasm, nucleus, or other organelle) are random (Baltes et al., 2017). The entry and integration of multiple copies of the gene of interest may also lead to undesirable effects such as gene silencing or altered gene expression (Rivera et al., 2014).

3.1.3.1. Biolistics particle bombardment

Biolistics, which is short for biological ballistics, is a method of nucleic acid delivery to cells by high-speed particle bombardment in a vacuum. The DNA or RNA-coated particles are propelled by a gene gun (see Figure 3.8), which is a pressurized apparatus, to transfect the cell samples (Carter and Shieh, 2015; Nature, 2019).

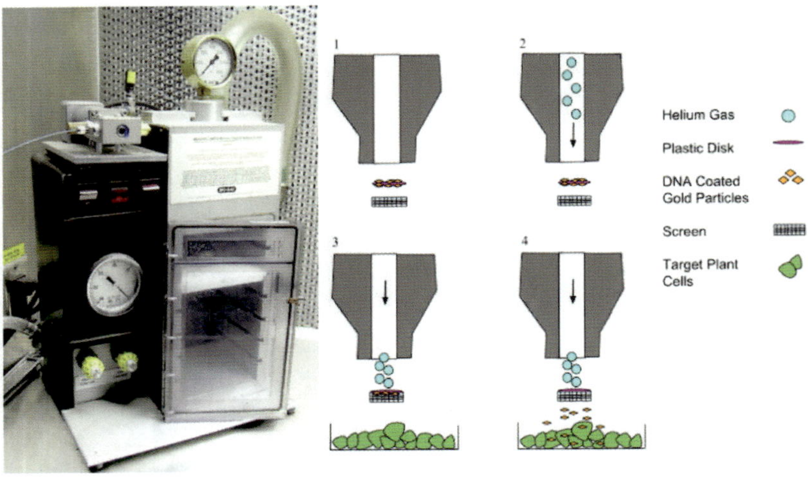

Figure 3.8: *A gene gun with a vacuum chamber (left image) is used to shoot target plant cells with DNA coated metal particles. (Right image) When it is turned on, helium flows through and moves the microprojectiles at accelerated speed through the plastic disk and screen and then injects the gene of interest into the plant cells.*

Source: Left image: Xmort, Public Domain; Right image: RachelBrooks15, Creative Commons License.

The first acceleration apparatus has a mechanism based on transferred mechanical impulse or a gunpowder charge and a stopping plate to accelerate tungsten particles to a speed of 328–656 meters per second (m/s). The second-generation instruments employ compressed gases to accelerate DNA-coated particles into plant cells (Christou, 1992). These pressurized gas-based devices allow for better control of bombardment parameter, distributes particles more uniformly over target cells, are gentler to the target cells, and yields higher transformation frequency (Hagio, 1998) (Figure 3.9).

Figure 3.9: *A handheld gene gun was developed which can inject DNA-coated particles into fully-grown plants.*

Source: https://home.cc.umanitoba.ca, Creative Commons License.

The negatively charged DNA associates with the metal particles to form DNA-metal complexes and these are layered uniformly inside Teflon-coated tubing, which are then cut into approximately 12.5 mm cartridges or "bullets." The cartridges are loaded into the gene gun and these should be used within 60 minutes. Gas pressure, usually from helium, builds up and provides the force to drive the particles down a shaft and to cause the rupture or plastic disk to break (see Figure 3.8). The DNA-metal complexes emerge from the filter screen and then penetrate the cell wall and cell membrane of the target plant cells (Carter and Shieh, 2015; Jinturkar et al., 2011). Once they reach the cytoplasm, the DNA is released from the particles and may be transiently expressed or integrated into the host genome for a more stable expression (Baltes et al., 2017).

A handheld gene gun was developed in 1996 which no longer requires a vacuum chamber (see Figure 3.9). Thus, the size of the target tissue is no longer limited by the chamber size and any target accessible to the barrel of the gun can be transformed (Hagio, 1998).

Introducing the gene of interest into the target cell by a gene gun overcomes barriers to genetic transformation such as the cell wall, certain cell surface receptors, and barriers in the nuclear envelope. Moreover, it reduces the likelihood of gene degradation since degradative lysosomal and endosomal enzymes are bypassed (Jinturkar et al., 2011). To enhance the efficiency of transformation, *A. tumefaciens*-mediated and biolistics methods can be combined. One way to do this is by wounding plant samples by bombarding them with particles prior to application of *A. tumefaciens* strains containing the gene of interest. This hybrid method was observed to enhance transformation frequency and stability (Bidney et al., 1992).

The main limitation of these devices is the requirement of embryogenic cultures, which can make the recovery of transgenic plants challenging. The uncontrolled acceleration can also excessively injure and kill the target cells. The depth of particle penetration is at random so that embryonic tissues and meristems are not ideal samples (Christou, 1992). Another disadvantage is the high cost of the gene gun and the metal (especially gold) particles (Jinturkar et al., 2011).

3.1.3.2. Electric discharge particle bombardment

Another mechanism for particle bombardment uses high voltage electric discharge into a droplet of water to generate shockwaves in order to propel the microprojectiles to the target tissue. Unlike the generally random nature of particle bombardment via the biolistics method, this can accurately control the location and depth of penetration so that majority of the DNA-coated particles are directed into the desired cell layer (Christou, 1992; McCabe and Christou, 1993). This technology is marketed as ACCELL, which stands for "Accessing any Cell" (National Museum of American History, 2019).

The apparatus bombards the target sample with DNA-coated particles via the force of an electric discharge. In one study, the discharge voltage used is 16 kilovolts (kV) and the vacuum pressure drawn in the bombardment chamber is 300-millimeter mercury (mm Hg) (Serres et al., 1992).

3.1.4. Gene Transfer by Electroporation

Electroporation is a process facilitated by short high-voltage electric impulses which induce the formation of transient pores in the cell membrane. As a result, plant samples take up macromolecules from the surrounding solution. For purposes of genetic transformation, this may be done using protoplasts or intact plant cells and tissues (National Academy of Sciences,

2004; Narusaka et al., 2012).

This technique is relatively simple, can be performed fast, and is not costly. On the other hand, it has low transformation efficiency and the procedure is labor-intensive (Rivera et al., 2014).

3.1.4.1. Electroporation of protoplasts

The thick plant cell wall is a barrier for DNA uptake and so the DNA of interest contained in a plasmid is added to a protoplast culture medium. Electric pulse is applied which temporarily destabilizes the cell membrane by creating pores, allowing the DNA to enter the protoplast cell (National Academy of Sciences, 2004). The protoplasts suspended in culture and the DNA is placed between electrodes in an electroporation cuvette. The cuvette is then placed in an enclosed discharge chamber (see Figure 3.10). Electric field is applied by a capacitor discharge (Rashid and Lateef, 2016; Narusaka et al., 2012). The strength of electrical field may range from 100–5,000 volts/centimeter (V/cm) while the duration of the pulse ranges from 0.01–100 millisecond (msec). These two electrical parameters are inversely related. Less rigorous electrical conditions are applied for stable transformation than for transient expression. Transformation efficiency generally increases with increased plasmid DNA concentration, which can reach up to 100 micrograms/milliliter (μg/mL). Linearizing the plasmid DNA and application of heat shock enhance stable transformation. On the other hand, addition of polyethylene glycol (PEG) stimulates transient expression and stable transformation (Joersbo and Brunstedt, 1996).

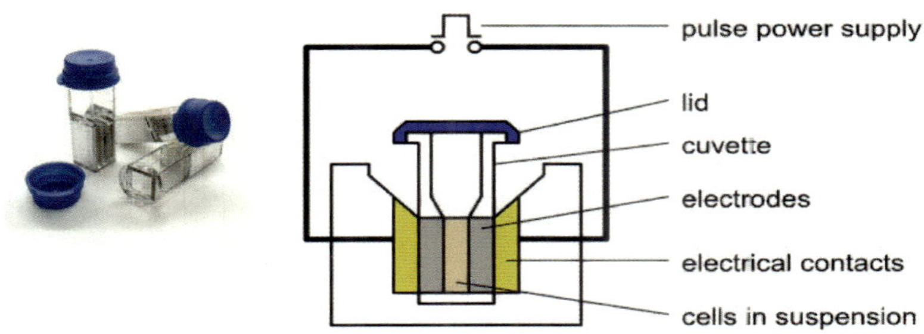

A B

Figure 3.10. *(A) Electroporation cuvettes are made up of plastic with aluminum electrodes and a lid. It can contain a maximum of 400 μL of DNA solution and cells or tissues in suspension. (B) The cuvette is then loaded into an electroporator with an electrical pulse power supply.*

Source: Richard Wheeler, Creative Commons License.

Transformation frequencies in electroporated protoplasts range from 0.0001–0.1%. On average, transgenic plants contain three copies of the gene of interest and all copies are usually integrated into the genome (Joersbo and Brunstedt, 1996). The protoplasts can then regenerate their cell walls and grow to whole, fertile transgenic plants. One major limitation of this type of electroporation is the poor efficiency of most plant species to regenerate from protoplasts (National Academy of Sciences, 2004). The viability of protoplasts and their capacity to divide are significantly reduced by both the electrical field and chemical substances applied to disorganize the cell walls (Rakoczy-Trojanowska, 2002). Moreover, preparing protoplasts adds difficulty to the technique.

3.1.4.2. Electroporation of intact plant cells and tissues

The principle of electroporation of intact plant cells and tissues is similar to that of electroporation of protoplasts. The samples used could either be pollens, microspores, leaf fragments, embryos, calli, seeds, or buds. The best results for plant transformation by electroporating cell suspensions were observed in maize. In one experiment, immature embryos and embryogenic calli were first briefly digested by pectolytic enzymes followed by electroporation. There were 90 transgenic plants that were regenerated from 1,440 embryos (i.e., 6.25% transformation efficiency). There were also 31 plants regenerated from 55 callus clusters (54.6%). These results were at par with the best results obtained through particle bombardment. In another study, the post-pulse addition of ascorbic acid is shown to significantly increase the transformation efficiency without any adverse effect on cell viability (Rakoczy-Trojanowska, 2002).

3.1.5. Silicon Carbide-Mediated Transformation (SCMT)

Silicon carbide-mediated transformation (SCMT) utilizes plant cells with intact cell walls. Small needle-type silicon carbide whiskers or fibers (an average of 0.6 μm in diameter and 10–80 μm in length) with very high tensile strength are mixed with plasmid DNA containing the gene of interest and plant cells. The mixture is vortexed and the resulting collision between the cells and the needle-like whiskers as well as intercellular collisions puncture the plant cell surface allowing DNA bound to the whiskers to enter the cells. The mixture is then plated on culture medium. The cultured cells are assessed whether there is successful DNA integration into the plant's genome by the expression of either a selectable marker or reporter gene or both.

Exposing the cells to high concentration of sorbitol increases the efficiency of DNA delivery by the silicon carbide whiskers (Rashid and Lateef, 2016; Kaeppler et al., 1992).

This method was used successfully in crops such as maize, tobacco, rice, and wheat. In one study involving soybeans, the transformation efficiency from SCMT is comparable to that of biolistics. In other cases, it has been used for wounding the cell to improve the efficiency of *A. tumefaciens*-mediated transformation. Its main advantages include the use of a variety of intact cells, the relatively simple and easy procedure, and the low equipment cost. However, it has its share of disadvantages like low transformation efficiency, cell damage, and the health hazard posed by inhalation of the fibers (Kaeppler et al., 1992; Darbani et al., 2008).

3.1.6. Microinjection of Gene of Interest

Microinjection is a method that introduces DNA into the nucleus or cytoplasm through a glass microcapillary-injection pipette and with the aid of a micromanipulator. Cells are immobilized with a holding pipette and gentle suction and then DNA, or even a whole chromosome, can be precisely introduced into the nucleus. Both the injection and holding pipettes contain mineral oil, which works as a cylinder (Rakoczy-Trojanowska, 2002).

This technique of gene transfer is usually used for animal cells since the glass micropipette struggles to penetrate the layer of lignin and cellulose in plant cell walls. Using protoplasts, like in the case of oil palms, instead of intact plant cells may resolve this difficulty (see Figure 3.11) but it increases the risk of releasing hydrolases and other toxic compounds stored in vacuoles to the cytoplasm. This may result to the death of the protoplasts. Vacuoles may be removed before microinjection, but this significantly reduces the capability for cell division and plant regeneration. Protoplasts are immobilized differently than intact cells; these are attached to glass by coating them with poly-L-lysine or agarose. However, poly-L-lysine can be toxic for some plant species and agarose reduces visibility in the area of microinjection (Rakoczy-Trojanowska, 2002). For these reasons, microinjection is not usually employed for plant genetic transformation, and if it is utilized, low transformation efficiency is observed. Moreover, this method requires expensive equipment, such as the micromanipulator, and involves tedious procedure. Nevertheless, transgenic plants were recovered via this method such as tobacco, petunia, rape, and barley (Rakoczy-Trojanowska, 2002; Darbani et al., 2008).

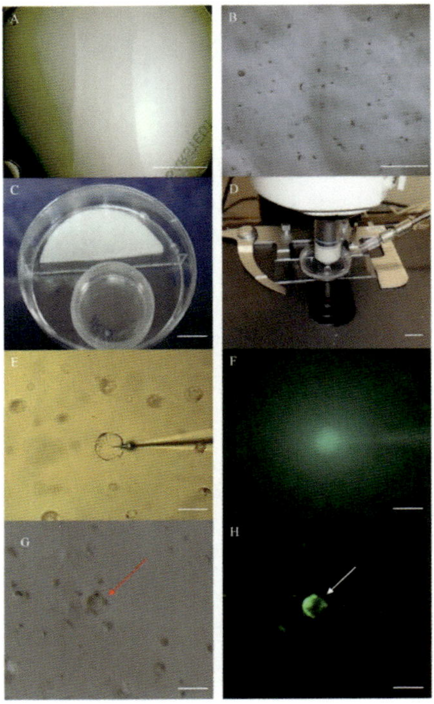

Figure 3.11: *(A) Protoplasts of oil palms are prepared and layered thinly onto a supporting medium. (B) The embedded protoplasts were observed under the microscope to confirm that these are arranged in a single planar layer. (C) The protoplasts were incubated in the dark for 3 days at 28°C. (D) These are then placed on a glass slide and observed under the microscope for DNA microinjection. (E) The DNA solution, which also contains a marker dye (Lucifer yellow) is injected into the protoplast. The Lucifer yellow allows the cytoplasm to be visualized under the microscope (F). (G and H) Green fluorescence protein (GFP) gene is incorporated into the plasmid DNA to serve as visual marker for both transient gene expression and stable expression after transgene integration. GFP fluorescence is detected in the cytoplasm of the injected protoplast (indicated by the arrow) after 3 days.*

Source: Masani, et al. (2014).

3.1.7. Genetic Transformation by Transfection

Transfection introduces DNA into protoplasts and has been used in the genetic manipulation of monocot crops. This technique is relatively simple and easy to perform with no expensive, specialized equipment needed. On the other hand, it has low transformation efficiency and, as previously pointed out, it is difficult to regenerate whole plants from protoplasts (Darbani et al., 2008).

3.1.7.1. Polymer-based transfection (Polyfection)

PEG, a hydrophilic long-chain polymer, is commonly used for introducing the gene of interest into protoplasts (Darbani et al., 2008). In one study, oil palm protoplasts were incubated with plasmid DNA and PEG for 10 min. The expression of the marker gene, which in this case is GFP, was observed after 72 hours indicating successful gene expression. However, transfection efficiency was low because most of the protoplasts were severely damaged and only a few of the fluorescent protoplasts survived. The addition of $MgCl_2$into the incubating solution increased the transfection efficiency significantly and higher concentration is directly related to increase in transfection efficiency (see Figure 3.12) (Masani et al., 2014).

Apart from oil palm, PEG-based transfection has been used to genetically transform rice protoplasts with up to 28% efficiency. In addition to $MgCl_2$, long-chain cations such as poly-L-ornithine, poly-L-lysine, or dextran sulfate as well as divalent cations like Zn^{2+} and Cu^{2+} were also observed to increase the efficiency of DNA uptake (Darbani et al., 2008).

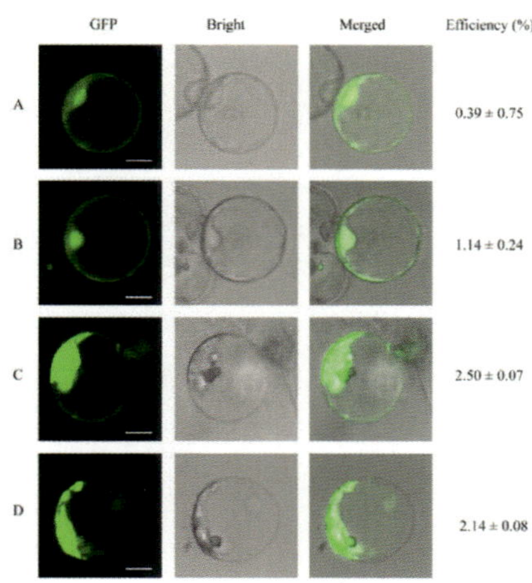

Figure 3.12: *Addition of increasing concentration of $MgCl_2$ with polyethylene glycol (PEG) increases transformation efficiency with green fluorescence protein (GFP) as the marker gene. (A) 10 millimolar (mM) resulted to 0.39% transformation efficiency, (B) 25 mM resulted in 1.14% efficiency, (C) 50 mM yielded 2.50% efficiency, and (D) 100 mM yielded 2.14% efficiency.*

Source: Masani et al. (2014).

3.1.7.2. Liposome-mediated transfection (Lipofection)

Lipofection uses cationic liposomes, spherical phospholipid bilayers with positively charged lipids, which interact with the negatively charged DNA as well as the negatively charged membranes of the target protoplasts. The gene of interest is encapsulated in a liposome. This is taken up by the protoplast cell possibly via endocytosis where the liposome fuses with the membrane of the target cell and its content is released into the cytosol. The gene of interest is then able to move into the nucleus and integrate into the host genome (Darbin et al., 2008). It is postulated that endocytosis of the liposome may also be mediated by other cellular factors such as clathrin or caveolin (see Figure 3.13) (Sarker and Takeoka, 2018).

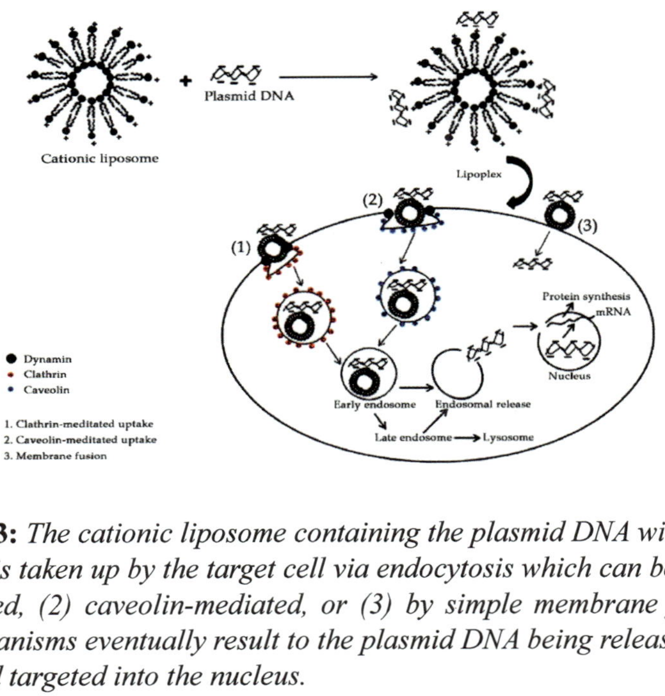

Figure 3.13: *The cationic liposome containing the plasmid DNA with the gene of interest is taken up by the target cell via endocytosis which can be (1) clathrin-mediated, (2) caveolin-mediated, or (3) by simple membrane fusion. All three mechanisms eventually result to the plasmid DNA being released into the cytosol and targeted into the nucleus.*

Source: Sarker and Takeoka (2018).

This technique is safe and simple to perform with readily available chemical reagents and it does not require expensive, sophisticated equipment. However, transformation efficiency is generally low with successful transgenic plants reported only in tobacco, wheat, and potato, so far (Darbin et al., 2008; Rao et al., 2016). Combining lipofection and PEG transfection resulted to higher transformation efficiency compared to the efficiency of each method done separately. Larger liposomes also yielded higher transfection efficiency than smaller ones (Darbin et al., 2008).

3.1.7.3. Peptide-mediated transfection

Protein transduction domains (PTDs) are small peptides with less than 20 amino acids, many of which are basic in nature. When these are cross-linked to macromolecules, they can facilitate the uptake of such macromolecules by target cells. Protein transduction was first used to deliver DNA into mammalian cells but was also done successfully in intact plants with the help of arginine-rich intracellular delivery (AID) peptide, a certain type of PTD (Chen et al., 2007) (Figure 3.14).

In one study involving mung bean and soybean, a plasmid DNA containing the gene of interest was directly delivered by AID peptide through the cell wall and membrane and into the nuclei. The gene of interest was expressed after 24 hours from delivery into the plant cells. The size of the peptide/DNA complex as well as the ratio between these two are determinants of efficient cellular uptake and subsequent integration into the genome. Overall, this method is simple, non-invasive, safe to perform, and efficient in terms of targeting the DNA into the nucleus (Chen et al., 2007).

3.1.8. Laser Microbeam-Mediated Transformation

Laser microbeam can be focused to puncture the cell wall and membrane creating transient self-healing holes that are about 0.5 μm in diameter. This allows the target cell to uptake the gene of interest contained in a plasmid vector (Darbin et al., 2008). The laser light can be manipulated and used as optical tweezers so that the transformation process is rapid, precise, and gentle to the target cell. This has been used to genetically transform wheat embryo-derived calli and melon hypocotyl explants. This requires an expensive laser system such as nitrogen lasers, excimer pumped dye lasers (see Figure 3.14) or titanium-sapphire lasers, the appropriate microscope, and a suitable computer program. In addition to the high equipment cost, this method is also laborious (Rivera et al., 2014; Badr et al., 2004; Khamis et al., 2009).

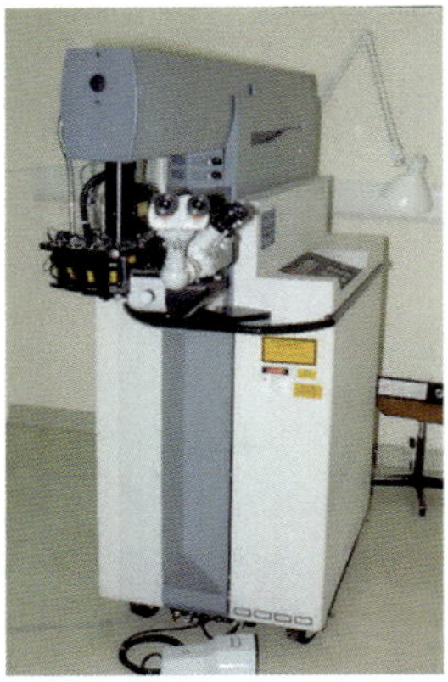

Figure 3.14: *An excimer laser produces a pseudo-molecule called excimer which generates ultraviolet laser light.*

Source: IoOALP, Creative Commons License.

3.1.9. Ultrasound-Mediated Genetic Transformation

The use of ultrasonic waves or ultrasound in introducing the gene of interest to the target plant is called sonication. In this method, acoustic waves temporarily change membrane permeability allowing the uptake of the gene of interest contained in a plasmid vector. This has been used in genetically manipulating animal cells and mostly plant protoplasts. There were reports of using this technique on intact plant cells and tissues, too. On the other hand, ultrasound can irreversibly break the membrane leading to cell death (Rivera et al., 2014) (Figure 3.15).

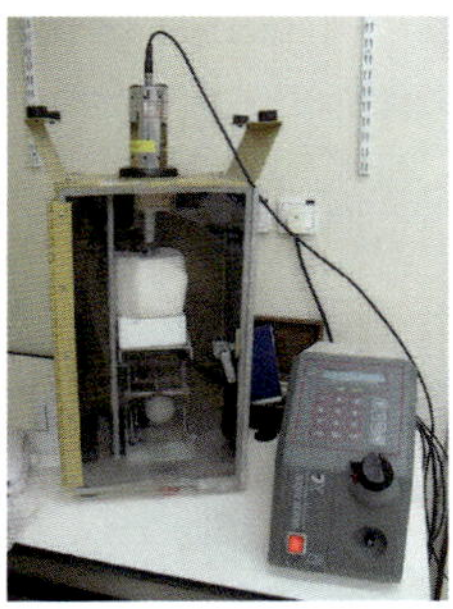

Figure 3.15: *A sonicator produces ultrasonic waves which can transiently puncture cell membranes enabling the entry of the gene of interest during transformation.*

Source: Eyal Bairey, Creative Commons.

Plant explants in a sonication medium and the plasmid vectors are placed together in a microcentrifuge tube. These are then exposed to ultrasound produced by a sonicator (see Figure 3.15) and transferred to fresh growth medium. In studies involving sugar beet and tobacco protoplasts, a brief exposure of the mixture to 20 kiloHertz (kHz) resulted in transient expression of the gene of interest. Stable transformation of tobacco leaf pieces was achieved by extending the duration of ultrasound exposure by 1,500–2,000 times (Darbin et al., 2008).

Sonication can also be used to improve the efficiency of viral vector-based transfection. It can also be used alongside *A. tumefaciens* in a process called Sonication-assisted *Agrobacterium*-mediated Transformation (SAAT). This method involves subjecting the plant tissue to ultrasonic waves for brief periods in the presence of *A. tumefaciens*. The ultrasound exposure induces several micro-wounds into the target plant cells so that *A. tumefaciens* infection becomes significantly more efficient and transient expression of the gene of interest is enhanced (Darbin et al., 2008).

3.2. APPLICATIONS OF GENETIC ENGINEERING IN PLANTS

Plants are the major food source of humans so that a decrease in crop production due to both biotic and abiotic stresses significantly impacts the population's well being. At the same time, crop yield must be increased to cater to the needs of a fast-growing human population. In addition, the ease and lower cost of plant propagation as opposed to animal reproduction makes plants ideal source of important biomolecules. In light of these facts, the main thrusts of genetic modification in plants are to make them more productive either as food source or bioreactors.

3.2.1. Genetic Modification for Resistance to Stresses

Plants are affected by biotic and abiotic stresses, which are due to both natural and man-made causes. Thus, there is increasing clamor for producing plants that can produce more yields with increasingly scarce resources such as water and land, survive the onslaught of plant pathogens, provide more nutrients, or synthesize biochemicals that are not inherently plant-based.

3.2.1.1. Resistance to biotic stresses

Biotic stresses come from other living organisms such as microorganisms (including viruses), insects, herbivores, or weeds which cause plant diseases, damage the plants (see Figure 3.16) or compete with the plants. These are traditionally counteracted by applying pesticides and herbicides which have negative consequences on the environment, the farmer, and the consumer. An alternative which can reduce the use of these chemicals is the introduction of genes in plants to make them more resistant to these biotic stresses (Low et al., 2018).

Figure 3.16: *The larva of the European corn borer (Otrinia nubilalis) (inset) is a major pest that damages the leaves, stalks, and ears of maize reducing crop yield.*

Source: Agronom, Creative Commons License; Inset: Keith Weller, Public Domain.

3.2.1.1.1. Insect-Resistant Transgenic Crops

The first batches of genetically modified crops that are insect resistant were commercialized in the mid-1990s. These include maize, potato, and cotton plants that harbor and express the ***Cry*** genes from *Bacillus thuringiensis*. The Cry proteins are endotoxins that can control insect pests, specifically lepidopteran larvae (caterpillars) (see Figure 3.16 inset). Upon ingestion by the insect, Cry proteins are solubilized in the midgut and cleaved at the

N-terminus to their truncated, active form. The active molecules bind to specific receptors in the midgut epithelial cells and insert into the membrane where they form pores. As a result, the cells die due to osmotic lysis and subsequently the insect dies (Gatehouse et al., 2011).

The early commercial varieties of *Bt* crops express only one type of Cry proteins. Examples are cotton and maize expressing Cry1Ac proteins with anti-lepidopteran activity. Eventually transgenic maize producing Cry3 proteins that are toxic to coleopteran pests, such as chrysomelid rootworms, were developed and are now commercially available (Gatehouse et al., 2011).

Eventually, other *Cry* genes were introduced and combined in a single crop, a process known as gene stacking. Examples of this include certain cotton varieties with Cry1F and Cry1Ac combination or Cry2Ab2 plus Cry1Ac. Combining different *Cry* genes in a single crop provides protection against a wider range of pests. Moreover, it can also prevent the development of resistance to toxin activity by the target pest/s. More recent varieties of transgenic maize express a combination of Cry proteins that are toxic to both lepidopteran and coleopteran pests. Another variety of transgenic maize expresses Vip3, a non-Cry *Bt* endotoxin, apart from the Cry proteins in order to provide resistance to multiple pests (Gatehouse et al., 2011; Gatehouse, 2008).

A variety of transgenic cotton has been produced which expresses Cry1Ac along with a modified cowpea trypsin inhibitor (CpTI). CpTI also has insecticidal properties but its main role is to reduce the possibility of target insects becoming resistant to this variety. Thus, this extends the effectivity of pest resistance (Gatehouse et al., 2011).

An important crop in the South and Southeast Asian countries is the eggplant. Eggplant growers suffer great losses in yield due to the eggplant fruit and shoot borer (EFSB) (see Figure 3.17). The female adult moth lays its eggs on the leaves. The eggs hatch and the larvae feed on leaf tissues and tunnel through the shoots and fruits. Transgenic eggplants with the *Cry1Ac* gene proved to be resistant against EFSB in studies conducted in Bangladesh, India, and the Philippines (ISAAA, 2018a; Shelton et al., 2018) (Figure 3.18).

Figure 3.17: *(A, B) Eggplant fruit and shoot borer (EFSB) damages the eggplant fruit by burrowing through it. (C) Farmers, often without personal protective equipment, traditionally combat this pest by spraying pesticide. (D, E right) Transgenic eggplants with Bacillus thuringiensis (Bt) toxins prove to be resistant to EFSB as opposed to non-transgenic eggplants (E left).*

Source: Anderson et al. (2019).

Figure 3.18: *The legume pod borer moth lays its eggs on the cowpea plant. When the caterpillars emerge, they feed on and subsequently damage the plant.*

Source: CSIRO, Creative Commons License.

Another important food crop, this time in Sub-Saharan Africa, is cowpea. The major pest of this crop is the legume pod borer which deposits eggs on the flower buds (see Figure 3.18). The larvae eventually destroy the flowers, young pods, and seeds greatly reducing the harvest. Transgenic cowpeas expressing Cry1Ab have proven to be resistant to legume pod borers, causing 100%, mortality after the caterpillars feed on the flower buds. Scientists are already evaluating the addition of Cry2Ab to the current variety of transgenic cowpeas in order to prevent the development of resistance by the pest to the toxins. It is estimated that the effectivity of a single *Cry* gene to resist the target insect is about 20 years, but two genes will sustain resistance for about 50 years (CSIRO, 2017; Isaac, 2018).

3.2.1.1.2. Non-Insect Pest Resistant Transgenic Crops

Insects are major pests of crops, but certain viruses, bacteria, fungi, and nematodes also destroy plants. This necessitates the development of plants that are resistant against these types of pests.

Banana and plantain, which are widely cultivated in Africa, Asia, Australia, and the Americas, can be infected with non-insect pests. One of these is the bacterium *Xanthomonas campestris* pv. *musacearum* which causes banana *Xanthomonas* wilt (BXW). It causes wilting of leaves or the male bud and premature fruit ripening that eventually leads to death of the plant and rotting of the fruit (see Figure 3.19). BXW causes big losses in the yield of bananas, particularly in East Africa. Two genes for resistance to BXW were sourced from sweet pepper namely, *hypersensitive response assisting protein* (*Hrap*) and *plant ferredoxin-like protein* (*Pflp*). These have been introduced either singly or as stacked genes to produce transgenic bananas that showed complete resistance to *X. campestris* pv. *musacearum*. Transgenic bananas expressing the rice pattern recognition receptor gene *Xa21* also demonstrated improved resistance against *X. campestris* pv. *musacearum* under glasshouse conditions (Conrow, 2017; Tripathi et al., 2017).

Figure 3.19: *Banana Xanthomonas wilt (BXW) is caused by Xanthomonas campestris pv. musacearum. It is characterized by leaf yellowing and wilting, exudation of bacterial ooze in the trunk, premature fruit ripening, and fruit discoloration.*

Source: Blomme et al. (2017).

Nematodes also cause significant damage to bananas. The major nematode pests in Sub-Saharan Africa include the migratory species *Radopholus similis*, *Pratylenchus goodeyi*, *Pratylenchus coffeae*, *Helicotylenchus multicinctus*, and the sedentary *Meloidogyne* spp. *R. similis* is considered to be the most damaging among these endoparasites causing extensive root necrosis as it migrates and feeds through the roots. This reduces root function and weakens plant anchorage so that the banana easily topples over during storms. *Pratylenchus* spp. damages the roots in the same manner as *R. similis*. *H. multicinctus* feeds in almost all growing areas but mainly in the root cortex causing some necrosis. The sedentary *Meloidogyne* spp. targets the roots (Tripathi et al., 2017).

Plantains that were genetically transformed to express cystatins exhibited improved nematode resistance in confined field trials. Cystatins, which are found in certain plants like maize, inhibit cysteine proteinases which are major digestive enzymes of many nematodes. Another possible source of resistance is a synthetic gene for a disulfide-constrained 7-mer peptide that reduces root invasion by the nematodes but does not kill them. The cystatin and the peptide have provided plantains resistance against *R. similis* and *H. multicinctus* as demonstrated by the failure of the nematodes to maintain their density on the growing plantain roots (Tripathi et al., 2017).

Another pest of banana, particularly the Cavendish variety, is the soil-borne fungus *Fusarium oxysporum* f.sp. *cubense* Tropical race 4 (FocTR4). This causes fusarium wilt or Panama disease characterized by extensive necrosis (i.e., yellow, wilting leaves and rotting trunks) as the fungi invade from the roots and then eventually enter the corm and pseudostem (see Figure 3.20). Plant death eventually ensues. This has affected Cavendish plantations in Asia, Australia, Africa, and even the Middle East. Two resistance genes were identified, one is the anti-apoptosis gene *Ced9* from the nematode *Caenorhabditis elegans*, while the second one is *resistance gene analog 2* (*RGA2*) isolated from *Musa acuminata* spp. *Malaccensis,* a wild variety of banana that has proven resistant to FocTR4. Transgenic Cavendish that expresses these genes was observed to be disease free. *RGA2* is innately present in Cavendish, but its expression is tenfold lower than the transgene form. One possibility is to enhance the expression of the endogenous *RGA2* through gene editing to provide non-transgenic resistance (Dale et al., 2017).

Figure 3.20: *(a, b) Wild-type Cavendish banana that is infected with Fusarium oxysporum* f.*sp. Cubense tropical race 4 (Foc TR4) has wilted, yellowing leaves and reddish-brown internal vascular discoloration. Transgenic lines expressing RGA2 (c, d) or Ced9 (e, f) proved to be resistant to the fungus.*

Source: Dale et al. (2017).

Maize streak disease (MSD) places a major constraint on maize production in Africa. This is caused by the geminivirus maize streak virus (MSV). The mechanism exploited to produce maize that shows resistance to MSV is called pathogen-derived resistance (PDR). In PDR, the plant is transformed with a portion of the viral genome to make it resistant to the virus from which the sequence is derived (Lomonossoff, 1995). Genes that could be used for PDR include those that code for the viral coat protein, replicases, or movement proteins. In some cases, these genes were mutated prior to introducing them to the target plant. It is proposed that if the host cell expresses dysfunctional pathogen-derived products, these could inhibit the pathogen by disrupting pathogen-host interactions (Sanford and Johnston, 1985). Apart from maize, PDR has been tried in food crops such as tomato, papaya, potato, cassava, sugarcane, barley, rice, and wheat. In creating a maize resistant to MSV, it was transformed with MSV which had mutated *replication-associated protein* (*rep*) gene. Transgenic plants showed a significant delay in the development of the symptoms (i.e., leaf necrosis and stunting), a decrease in symptom severity, and higher survival rates than non-transgenic plants (Shepherd et al., 2007). To further improve the transgenic maize, the transgene expression system has been made inducible. This means that the mutant *rep* transgene is expressed only in MSV-infected cells. This was accomplished by splitting the transgene so that it would be translated only in the presence of an infecting MSV's replication associated protein (Shepherd et al., 2014).

3.2.1.2. Resistance to abiotic stresses

Abiotic stresses stem from non-living factors that can deplete the productivity of food crops. Common examples of these are herbicide exposure, drought, floods, increasing soil salinity, and cold temperature. There are common genes involved in plant response to abiotic stresses. This means that genes involved in conferring tolerance to drought, for example, also play a role in salinity or cold tolerance.

3.2.1.2.1. Herbicide Resistant Crops

Weeds pose major threats to crops by competing for space and nutrients. These are usually managed by spraying herbicides. While there are selective herbicides, broad-spectrum ones are preferred in order to target a wide range of weeds. However, these chemicals damage the crops posing a complication to weed management. In the early 1970s, herbicide-resistant weeds were discovered and became sources of genes for developing herbicide-resistant crops. The early versions of these crops were produced via traditional breeding. Eventually, genetic engineering methods were employed to generate transgenic herbicide-resistant crops (Madsen and Streibig, 2019).

The first batch of herbicide-resistant transgenic crops were those that are resistant to the herbicide glyphosate. Glyphosate inhibits the enzyme 5-enolpyruvylshikimate 3-phosphate synthetase (EPSPS). This is an enzyme involved in the shikimic acid pathway, which is indirectly involved in the synthesis of the essential amino acids phenylalanine, tyrosine, and tryptophan. Since this pathway is generally present and necessary in plants, glyphosate is a broad-spectrum herbicide affecting a wide range of plants.

However, since the pathway is not found in animals and humans, it has minimal effect on them (Gupta and Crissman, 2013).

There are two genes that code for a glyphosate-insensitive EPSPS which are used in commercially available transgenic crops that are glyphosate-resistant. One is *cp4 epsps* gene from *A. tumefaciens* strain CP4. The other one is a mutated *zm-2mepsps* from maize. One possible disadvantage of using these in producing herbicide-resistant crops is that the glyphosate remains in the plant and accumulates in meristems, where it may interfere with reproductive development and may decrease yield. Another gene of interest is from the bacterium *Ochrobactrum anthropic* strain LBAA that codes for glyphosate oxidoreductase (GOX). This enzyme detoxifies or degrades glyphosate. The benefit of glyphosate detoxification is that the breakdown and removal of the herbicide residue enhance herbicide tolerance and allow spraying during the reproductive stage (Green and Owen, 2011; Pollegioni et al., 2011).

One of the most widely used glyphosate herbicide is Roundup, produced by the biotech company Monsanto. Then in 1996, the same company released transgenic herbicide-resistant soybeans with the brand name Roundup Ready. This was followed by Roundup Ready maize and cotton. The use of these crops reduced the use of toxic pesticides, and allowed farmers to adopt direct seeding with no or reduced tilling. Tillage or manual soil cultivation has many benefits including the mechanical destruction of weeds prior to the next planting season. However, doing so for the entire farming area is quite tedious. Not surprisingly, Roundup Ready crops proved to be popular among farmers (Wilkerson, 2015).

On the other hand, using only one herbicide may induce weeds to mutate and develop resistance to it, which they can pass on to their offspring. Thus, the use of Roundup and Roundup Ready crops should be supplemented by other weed management practices. One strategy that can be employed to reduce the probability of weeds developing resistance is to alternate the use of Roundup with another herbicide that kills weeds using a different mechanism. This makes it more difficult for resistance to develop since two mutations are needed. However, using another herbicide would destroy the Roundup Ready crops, too making this approach unpopular to farmers (Wilkerson, 2015).

Another approach is to do crop rotation, i.e., change the crop planted on the land each year. These different crops use different nutrients and root into the soil differently making the environment less stable and less habitable for weeds. Since this entails more work for farmers, this is rarely observed (Wilkerson, 2015).

As a result of relying only on glyphosate for weed management, there are already reports of super weeds or weeds that have developed resistance to glyphosate. As a result, many farmers are spraying more glyphosate and other herbicides to combat these weeds. On the biotechnology front, transgenic plants that are resistant to additional herbicides have been developed. Transgenic soybeans and cottons with glyphosate-resistance transgenes are transformed with another gene for resistance to 2,4-D, an herbicide that controls broadleaf weeds. Monsanto also released soybeans with resistance to the herbicides glyphosate and dicamba. Both 2,4-D and dicamba are volatile which may drift into neighboring fields and damage other plants. In 2017, many farmers in the US complained of crop damage due to dicamba drift prompting the states of Arkansas and Missouri to ban dicamba spraying for some of the growing season (Wilkerson, 2015; InTeGrate, 2018).

Apart from glyphosate-resistant crops, there are other transgenic crops which are resistant to other herbicides. Examples are glufosinate-resistant soybeans, cotton, and jute (see Figure 3.21). Glufosinate is a broad-spectrum herbicide but compared to glyphosate, it is more effective against annual broadleaf (dicot) weeds than grasses (monocot). Its active ingredient, L-phosphinothricin, inhibits glutamine synthetase (GS) involved in nitrogen assimilation (AgFax.com, 2010; Carbonari et al., 2016) and detoxification of ammonia (ISAAA, 2018b). GS inhibition leads to glutamine deficiency, glutamate accumulation, and highly toxic ammonia build up (Carbonari et al., 2016). Genes that confer resistance against glufosinate are PAT (*phosphinothricin acetyltransferase*) and its homologue, *bar* (*bialaphos resistance*), from *Streptomyces viridochromogenes* and *S. hygroscopicus*. These encode *N*-acetyl transferases that inactivate glufosinate by acetylating L-phosphinothricin into *N*-acetyl-L-glufosinate ammonium which does not inhibit GS (Green and Owen, 2011; Majumder et al., 2018; Carbonari et al., 2016).

Figure 3.21: *First generation of transgenic lines (T_1) of jute (JBC1, JBC2, JBC3) with the bar gene were grown and then sprayed with the glufosinate herbicide Basta®. (Upper row) Non-transgenic, wild-type (WT) plants dies 6 days after the herbicide application while the transgenic plants recovered from the herbicide stress and grew new leaves and grew longer. (Lower row) Plants of T_2 generation also showed tolerance to Basta® application as opposed to the WT plants.*

Source: Majumder et al. (2018).

Glyphosate and glufosinate are the most widely used commercial herbicides but there are other types of herbicides with corresponding genes for tolerating them. Examples of these are the *aad-1* and *aad-12*genes that confer tolerance to 2,4-Dicholorophenoxyacetic acid, *hppd* for isoxaflutole, *bxn* for oxynil, and *als* for sulfonylurea herbicide (Majumder et al., 2018).

3.2.1.2.2. .Drought-Tolerant Crops

Drought greatly affects crops due to inadequate water or moisture in the soil, which may be due to insufficient rainfall. Most crops are adversely affected by drought during the flowering to seed development stage. It accounts for about 70% of potential crop yield loses worldwide. Therefore, efficient water management and development of drought-resistant crops are important strategies to meet food demands with less water (Mitra, 2001; ISAAA, 2008).

Drought resistance is a trait that enables plants to thrive in a water-deficient environment with minimum loss in terms of their yield. There are various mechanisms involved in drought resistance, which include drought escape, drought avoidance, and drought tolerance (Mitra, 2001). Drought tolerant varieties of crops are developed by either conventional or genetic engineering methods. Using genetic engineering methods to produce drought tolerant plants requires the identification of genes involved in stress tolerance in plants prior to inserting them into the target plants (ISAAA, 2008).

Drought triggers a variety of physiological responses in plants, which involve several hundred genes. Their expression is either induced or repressed under drought condition. When water is insufficient, the physiological effects on the plant include leaf wilting, leaf area reduction, leaf abscission, and stimulation of root growth due to nutrients being directed to the roots. Since plant resources are deviated to support root growth during drought, plants are more susceptible to water scarcity during the reproductive stages when flowers and seed develop. Drought also induces the production of abscisic acid (ABA), a plant stress hormone, which causes the leaf stomates to close in order to minimize water loss via transpiration. However, this decreases the rate of photosynthesis. This response is reversible, i.e., in the presence of water, the level of ABA decreases causing the stomates to re-open. Certain stress response regulatory pathways are ABA-dependent while there are those which are independent of ABA (ISAAA, 2008).

Plant cells need to maintain water-solute balance relative to their surroundings. When the cells contain more solutes such as sugars, amino acids, organic acids, and ions compared to their environment, they absorb water. The presence of ions in the cell inhibits certain cellular enzymes so these must be removed from the cytosol and stored in the vacuoles. On the other hand, compatible solutes such as sugar alcohols, the amino acid proline, and glycine betaine do not inhibit enzymes. Thus, the synthesis of these compounds enhances drought tolerance (ISAAA, 2008).

Drought tolerance involves multiple genes that are activated in response to drought. These are involved in the perception and transmission of drought stress signal (see Figure 3.22). Some of these genes code for proteins that protect the cells from desiccation. These include membrane transporters, ion channels, receptor-like kinases, and calcium-binding proteins involved in accumulation of compatible solutes, passive cellular transport, and water transport. Other genes encode proteins that facilitate the closure of stomates and protection of the cell from damage by reactive oxygen species (ROS). Another set of genes encode regulatory proteins that control stress signal transduction and gene expression such as transcription factors (TFs) like *AP2*, *bHLH*, *bZIP*, *MYB*, *MADS*, *NAC*, and *WRKY* (ISAAA, 2008; Gupta et al., 2017).

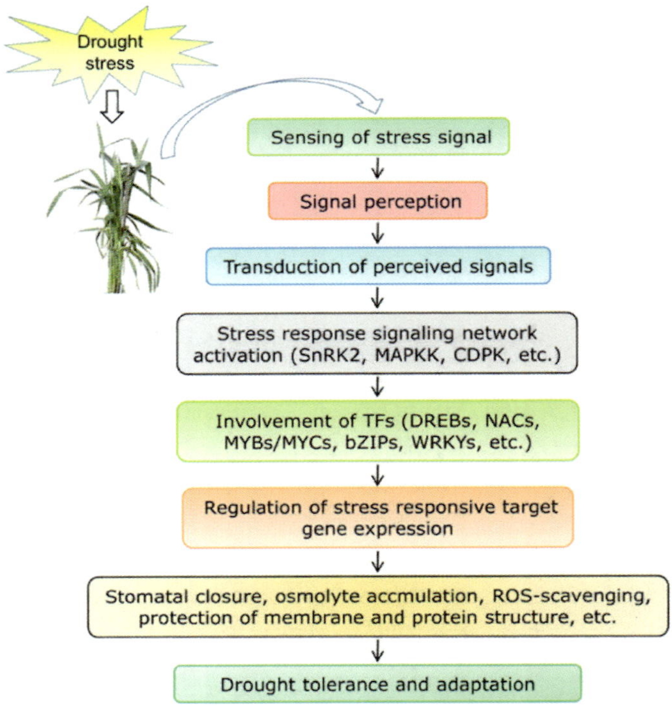

Figure 3.22: *Several steps are involved in the expression of stress tolerance, beginning from the perception of the drought stress and transducing the signal to transcription factors (TFs) for the expression of the genes involved in adaptation. Thus, drought tolerance is a complex phenotype and it is polygenic.*

Source: Gupta, Balyan, and Gahlaut (2017).

Genes that are involved in maintaining cellular structure and functions are also affected by limited water availability making them possible targets for developing drought tolerant crops. These include genes involved in cell growth that are mostly downregulated under drought conditions, as well as genes for hormone synthesis (such as ABA), biosynthesis of compatible solutes (such as proline), and carbohydrate metabolism, which are upregulated. There are also genes expressed exclusively in the roots under drought conditions such as those that encode expansins, extensins, xyloglucans, and cellulose synthase. Other genes code for protective proteins such as late embryogenesis abundant (LEA) proteins and chaperons or heat shock proteins. Finally, genes encoding isopentenyl transferase (IPT) delay senescence, which enables plants to maintain high photosynthetic activity during episodes of drought (Gupta et al., 2017).

A. thaliana served as the model plant in identifying the genes and elucidating the basic pathways involved in a plant's response to stress like drought, cold, and salinity. *A. thaliana* is not a food crop but the information about genes involved in stress tolerance derived from its genetic manipulation is useful in improving food crops (ISAAA, 2008).

One gene identified to be important in drought tolerance in *A. thaliana* is *ERA1*, which encodes the β-subunit of farnesyl-transferase (AtFTB) and is involved in ABA sensing. Plants that lack *ERA1* exhibit increased sensitivity to ABA and thus, show reduced wilting during drought conditions. This indicates that a certain ABA modulator must be farnesylated in order to downregulate ABA sensitivity in *A. thaliana*. The absence of *ERA1* also results to altered phenotypes such as delayed senescence, altered apical and floral meristem structures, and severe reduction in seed yield (ISAAA, 2008; Wang et al., 2005).

A transgenic line of canola was developed with downregulated levels of AtFTB, and therefore hypersensitive to ABA. This was accomplished by introducing both *ERA1* sense and antisense constructs into the target plant. Additionally, the expression of the transgenic *ERA1* antisense construct is driven by the *rd29A* promoter, which initiates expression only under drought condition. Under drought stress conditions, the transgenic canola showed enhanced ABA sensitivity, significantly reduced stomatal conductance and water transpiration, and greater resistance to seed abortion during flowering as compared to the non-transgenic lines. Field trials also show that transgenic canola had significantly higher seed yield than non-transgenic ones under moderate drought stress conditions. Under the *rd29A* promoter, the observed downregulation of farnesyl transferase in the transgenic canola is a conditional and reversible process, which depends on the soil water levels (Wang et al., 2005).

Another gene of interest is for 9-*cis*-epoxycarotenoid dioxygenase (NCED), a key enzyme in ABA biosynthesis. Its overexpression increased ABA levels and induced the transcription of drought- and ABA-inducible genes. These lead to reduced leaf transpiration rates and improved drought tolerance in transgenic *A. thaliana* (Iuchi et al., 2001). On the other hand, *CYP707A3* belongs to a cytochrome P450 family that catalyzes the degradation of ABA. Introducing a mutant form of this gene to *A. thaliana* resulted to higher ABA levels and more enhanced tolerance under drought conditions compared to the wild type (Umezawa et al., 2006).

Drought responsive element binding 1 (*DREB1*) and *DREB2* genes are also discovered to increase drought tolerance in *A. thaliana.* These are TFs that are essential in ABA-independent stress response pathways. These bind the promoters of drought-, high salinity-, and cold-inducible genes, particularly the DRE/CRT (dehydration-responsive element/C-RepeaT). The overexpression of these two TFs confers tolerance of transgenic *A. thaliana* plants to drought, salt stress, and freezing. Homologues of these two genes are present in rice and overexpression of *DREB1* in rice also improved its drought and cold tolerance (Shinozaki and Yamaguchi-Shinozaki, 2007). The presence of these two TFs and their role in stress tolerance have been reported in other crops such as tomato, barley, canola, maize, soybean, rye, and wheat indicating that these are involved in a conserved, universal stress defense pathway in plants (ISAAA, 2008).

The drought-tolerant maize variety developed by Monsanto contains the *cold shock protein B* (*cspB*) gene from *Bacillus subtilis.* Cold shock proteins function as RNA chaperones, binding and unfolding tangled RNA molecules so that they can retain their function during the onset of stress (DiLeo, 2012).

In rice, the gene *Oryza sativa ethylene responsive binding protein 1 (OsEREBP1)* was found to confer tolerance to biotic and abiotic stresses, which include drought. It codes for an AP2/ERF transcription factor, which is part of a signal transduction pathway in a specific interaction between rice and its bacterial pathogen, *Xanthomonas oryzae* pv. *oryzae* (Xoo). Rice plants were genetically transformed with *OsEREBP1* driven by a maize ubiquitin promoter in order to induce gene overexpression. Compared to the non-transgenic plants, the transgenic ones exhibited drought tolerance during the vegetative stage (see Figure 3.23A). Specifically, they showed delay in stress-induced symptoms such as leaf-rolling and wilting. At the reproductive stage, drought stress was followed by rewatering to observe recovery of the plants. The transgenic plants had more tillers, which eventually formed into panicles and grains (see Figure 3.23B) (Jisha et al., 2015).

Figure 3.23: *(A) Transgenic rice plants (4–3 and 2–4) overexpressing Os-EREBP1 showed delayed onset of leaf rolling and wilting when subjected to drought stress compared to the non-transgenic plants (Kit). (B) Drought was imposed at the reproductive stage by withdrawing water during the panicle initiation stage until the plants were completely brown before they were rewatered. The transgenic rice plants showed new tillers which formed into panicles and grains indicating that they recovered from drought.*

Source: Jisha et al. (2015).

3.2.1.2.3. Submergence-Tolerant Crops

Millions of farmers lose their entire crops to flooding annually. Most of these live in the poorest areas of the world. Most dryland cereals, such as maize, wheat, and barley are sensitive to waterlogging causing 20%–40% yield loss. Waterlogging inhibits root growth and function because of shortages in oxygen, which restrict root respiration. In addition, the concentration of potentially toxic compounds increases in flooded, anoxic soils. These compounds can enter the plant through the roots, damaging both root and shoot tissues. When the soil is reaerated, the ethanol trapped in the plant tissues is converted to acetaldehyde, causing further cell injuries. Excessive accumulation of ROS also occurs upon oxygen deprivation or re-oxygenation under light conditions (Ismail, 2018).

One crop greatly affected by flooding is rice. Although 24% of the rice crop is grown in paddies, most rice varieties will die if fully submerged for more than three days (Ronald, 2019). One gene of interest for improving survival after submergence is *pdc1*, which codes for pyruvate decarboxylase (PDC). Rice lines were transformed with rice *pdc1* resulting to overexpression of the gene and higher levels of PDC. This enzyme is involved in alcohol fermentation. Increased PDC activity enhances metabolic capacity of the plant under anaerobic conditions when submerged and thus, increases its tolerance to submergence (Quimio et al., 2000).

A few cultivated varieties (cultivars), such as the *Oryza sativa* ssp. *indica* cultivar FR13, can survive up to two weeks of complete submergence. This is due to major quantitative trait locus (QTL) called *Submergence 1* (*Sub1*) located near the centromere of chromosome 9. A QTL is a genomic region responsible for the variation of a quantitative trait. There are three genes at the *Sub1* locus namely, *Sub1A, Sub1B,* and *Sub1C*. These possibly code for ethylene response factors (ERF). A submergent-intolerant cultivar, *O. sativa* ssp. *japonica*, was genetically transformed with *Sub1A* under the control of a promoter that induces overexpression of the gene. The transgenic lines showed enhance submergence tolerance (Xu et al., 2006). Better results were observed when non-tolerant rice cultivars were introgressed with *Sub1* via marker assisted breeding (MAS).

Overexpression of *OsEREBP1* was previously discussed to confer drought tolerance to rice. In addition, this also enhances submergence tolerance in rice. This is possibly because the gene codes for a transcription factor that upregulates *Adh1*, the gene for alcohol dehydrogenase (ADH). ADH, just like PDC, is part of the alcohol fermentation pathway, which provides energy during submerged conditions. *Sub1A* has been noted to also upregulate *Adh1* expression (Jisha et al., 2015) (Figure 3.24).

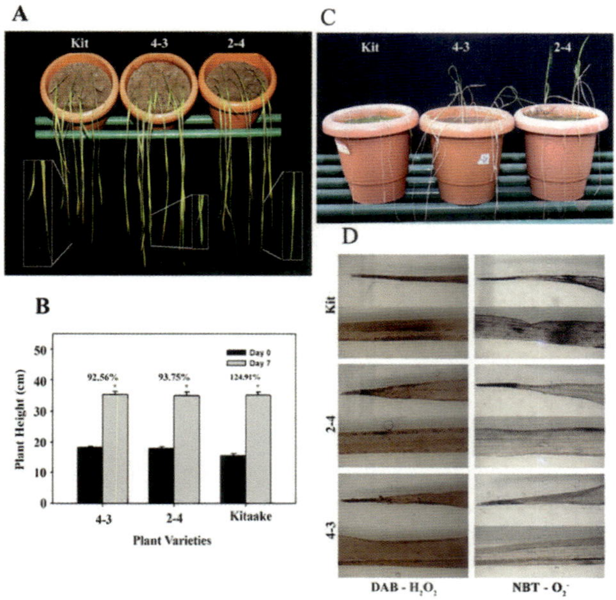

Figure 3.24: *(A) After being subjected to submergence stress, transgenic rice plants (4-3, 2-4), which overexpress OsEREBP1, showed flat and green leaves as opposed to the non-transgenic lines (Kit) which had rolled and yellowed leaves. (B) When compared to the transgenic lines, the non-transgenic, Kitaake lines showed higher length difference after submergence. (C) Transgenic lines were able to recover after prolonged submergence as indicated by the growth of new leaves, which were not observed in the non-transgenic lines. (D) After submergence, the leaves were stained with 3-3'diaminobenzidine tetrahydrochloride (DAB) and nitro-blue tetrazolium (NBT) to visualize H_2O_2 and O_2^- accumulation, respectively. The intensity of brown and blue colors was greater in the non-transgenic lines compared to the transgenic ones.*

Source: Jisha et al. (2015).

Rice plants were made to overexpress *OsEREBP1* and the resulting transgenic plants were compared to their non-transgenic counterpart after submergent conditions. Upon desubmergence, the transgenic rice plants had flat and green upper leaves while the non-transgenic ones had severe leaf rolling and yellowing (see Figure 3.24A). The rate of plant growth can also be an indicator of tolerance to submergence since submergence causes leaf and internode elongation and decreases plant growth. Leaf elongation was greater in the non-transgenic plants than in transgenic rice (see Figure 3.24B). Moreover, the growth rate of non-transgenic plants was significantly lower when compared to those that were not subjected to submergence. After a more prolonged submergence, most of the transgenic lines recovered and produced new leaves as opposed to the non-transgenic lines whose withered and dried leaves failed to recover (see Figure 3.24C). Sudden re-exposure to air during desubmergence can lead to excessive ROS generation which can be damaging to the plant. The non-transgenic lines had higher levels of superoxide O_2^- and H_2O_2 (see Figure 3.24D). All these indicate that *OsEREBP1* enhances submergence tolerance and recovery (Jisha et al., 2015).

Drought-tolerant transgenic tobacco was also developed by introducing the *sacB* gene from *Bacillus subtilis*. This gene codes for Levan sucrase which is an enzyme for fructan biosynthesis. Fructans are polyfructose molecules which function as the main storage carbohydrate of certain plant species. Due to their solubility, they may also play a role in helping plants survive osmotic stress brought about by drought or cold temperature. This was proven by the transgenic plants exhibiting tolerance to drought treatment as indicated by higher growth rate, fresh weight (FW), and dry weight compared to the wild type (Pilon-Smits et al., 1995).

3.2.1.2.4. Salinity-Tolerant Crops

A saline soil is one in which the electrical conductivity (EC) of the saturation extract (EC_e) in the root zone exceeds approximately 40 millimolar (mM) NaCl at 25°C and has an exchangeable sodium of 15%. It is estimated that 20% of total cultivated and 33% of irrigated agricultural lands have high salinity. These areas are increasing at a rate of 10% per year due to low precipitation, high surface evaporation, weathering of native rocks, irrigation with saline water, and poor cultural practices (Shrivastava and Kumar, 2015).

Soil salinity is a major abiotic stress that negatively affects plant growth rates, tillering, and reproductive development. Within minutes of salt application, the plants close their stomates, their leaf temperature increases, and shoot elongation is inhibited. These rapid responses are independent of the accumulation of salts in the shoot. Thus, these are termed 'shoot salt accumulation independent effects.' Over several days to weeks from exposure to salt stress, there is slower onset of growth inhibition especially in the older leaves, causing the premature senescence of these leaves. This is called 'ionic phase' of salt toxicity, which is due to both salt accumulation and the inability of the shoot to tolerate the toxic amounts of salt (Roy et al., 2014).

There are three main categories of mechanisms employed by plants to tolerate salt stress (see Figure 3.25). First is osmotic tolerance, which is triggered before the accumulation of Na^+ in the shoot and is regulated by long distance signals that reduce shoot growth. Another mechanism is ion exclusion where Na^+ and Cl^- transport processes in the roots reduce the accumulation of toxic concentrations of such ions in the leaves. These include retrieval of Na^+ from the xylem, compartmentalization of ions in the vacuoles of cortical cells, and efflux of ions back to the soil. The third mechanism is tissue tolerance, in which high salt concentrations are found in the leaves but are compartmentalized in vacuoles inside the cells. This requires ion transporters, proton pumps, and compatible solute synthesis (Roy et al., 2014).

Considering that several processes are involved in conferring salinity tolerance to plants, there are several genes that can be used to genetically transform crops to survive salt stress. The *high affinity potassium transporter* (*HKT*) gene family and the salt overly sensitive (SOS) pathway may be involved in the ion exclusion mechanism of responding to salinity owing to their important role in regulating Na^+ transport within the plant. *SOS1* gene that codes for a plasma membrane Na^+/H^+ antiporter was isolated from cotton and was introduced into *A. thaliana*. Its overexpression in the transgenic *A. thaliana* enhanced tolerance to salt stress as indicated by their lower malondialdehyde (MDA) content, which is a marker of oxidative lipid injury caused by environmental stress, decreased Na^+/K^+ ratio, and better survival under salt treatment (see Figure 3.25) (Chen et al., 2017; Kong et al., 2016).

Incorporating *HKT* alleles from *Triticum monococcum* to durum wheat via MAS successfully improved the plant's salinity tolerance. Producing transgenic lines using the same genes was moderately successful (Roy et al., 2014). For example, barley was genetically transformed by overexpressing *HKT2:1*, a gene for Na^+ transporter. The transgenic barley showed enhanced Na^+ uptake, higher Na^+ concentrations in the xylem sap, and enhanced translocation of Na^+ to leaves when the plant was grown in the presence of high NaCl concentration. While barley is a relatively salt-tolerant crop, overexpression of the transgene reinforced its salt-including behavior (Mian et al., 2011). Another set of genes for regulating Na^+ and K^+ transport is *halotolerance protein* (*HAL*). *HAL1* was introduced into tomato and the resulting progeny of the transgenic lines with one or four copies of the transgene showed higher level of tolerance to salinity stress. In addition, these were able to retain more K^+ than the non-transgenic plants under salt stress (Gisbert et al., 2000).

Figure 3.25: *Transgenic lines of Arabidopsis thaliana (L12, L14), which over-express SOS1 gene from cotton, survived salt stress treatment (250 mM NaCl) for 7 days as opposed to the wild type plants (WT).*

Source: Chen et al. (2017).

In terms of the tissue tolerance mechanism, *NHX*, which codes for vacuolar Na^+/H^+ antiporters, and genes for vacuolar H^+ pyrophosphatases (e.g., *AVP1*), which are involved in compatible solute synthesis, are possible candidate genes for genetically engineering salinity tolerance among crops. Buckwheat and rapeseed that overexpress *NHX* from *Arabidopsis* and tomato that overexpresses *AVP1* from rice demonstrate salinity tolerance (Roy et al., 2014; Zhang et al., 2001).

Another mechanism for tissue tolerance involves detoxification enzymes such as glutathione-S-transferases (GSTs). These enzymes can metabolize various toxic exogenous compounds, hence are commonly known for their role in herbicide detoxification. These also function as non-enzymatic carriers in intracellular transport allowing transport into vacuoles via glutathione pumps. In addition, plant GSTs function as peroxidases and may play a role in enhancing tolerance to abiotic stresses such as cold, drought, and salinity. A GST gene from tomato, *LeGSTU2*, was introduced into *A. thaliana*. The resulting transgenic plants showed significantly higher antioxidant activity which correlated to enhanced tolerance to both salinity and drought stress (Xu et al., 2015) (see Figure 3.26).

Figure 3.26: *Three-week old transgenic (OE3, OE7, OE10) and wild type (WT) Arabidopsis thaliana plants were subjected to salt and drought stress conditions for 14 days. Results showed that significantly more transgenic plants survived than the WT.*

Source: Xu et al. (2015).

3.2.1.2.5. Cold-Tolerant Crops

Plants vary in their response to chilling (0–15°C) and freezing (<0°C) temperatures. Generally, plants from temperate regions are tolerant to chilling at variable degrees. Their tolerance can be enhanced by cold acclimation, a process in which they are exposed to chilling, non-freezing temperatures. On the other hand, plants from tropical and subtropical regions are sensitive to chilling stress and lack the cold acclimation mechanism. Severe dehydration is associated with freezing and this may result to membrane damage. Thus, these plants are injured by low temperatures characterized by symptoms like chlorosis, necrosis, or growth retardation. These may eventually lead to the death of the plant. Cold stress could therefore affect crop productivity (Sanghera et al., 2011).

Tolerance to low temperature is a complex trait that involves several cellular parts and metabolic pathways. It involves increased chlorophyll accumulation, reduced sensitivity of photosynthesis, and improved germination, pollen fertility, and seed set. It is polygenic and the mechanisms of how these multiple genes interact and control cold tolerance are not yet fully elucidated (Sanghera et al., 2011).

There are two groups of genes that are identified to be associated with cold stress tolerance. The first group is stress-inducible genes that code for chaperones, LEA proteins, osmotin, anti-freeze proteins, mRNA-binding proteins, key enzymes for osmolyte biosynthesis (e.g., proline, water channels, sugar, and proline transporters, detoxification enzymes), fatty acid metabolism enzymes, and lipid-transfer proteins. The second group includes protein factors involved in regulating signal transduction and gene expression such as TFs that control different stress-inducible genes. Some of these genes have been identified and these were used to generate transgenic lines of different crops with improved cold stress tolerance. Table 3.1 presents some of these genes that were introduced into various crops such as tobacco, rice, alfalfa, and tomato. Several of these genes also confer tolerance to drought and salinity (Sanghera et al., 2011).

Table 3.1: *A Selection of Genes Used in Producing Cold Stress-Tolerant Transgenic Crop Plants*

Gene/s (Gene Product)	Function	Transgenic Host Plant	Performance of Transgenic Plants	References
gpat (Glycerol 3-phosphate acyltransferase) from *Arabidopsis thaliana*	Unsaturation of fatty acid	Tobacco	Less chilling damage to photosynthetic activity and to the leaves than the wild type	Murata et al., 1992
sod (Cu/Zn superoxide dismutase)from pea	Dismutation of toxic reactive oxygen intermediate as a component of the active oxygen scavenging system of chloroplasts	Tobacco	20% higher photosynthetic activity during chilling than the untransformed plants	Gupta et al., 1995
mn-sod (Mn-superoxide dismutase) from *Nicotiana plumbaginifolia* (Tex-Mex tobacco)	Dismutation of reactive oxygen intermediates in the mitochondria as part of mechanisms to detoxify oxygen radicals	Alfalfa	Increased winter survival during field trials	McKersie et al., 1996

GST (Glutathione S-transferase) and *GPX* (Glutathione peroxidase)	Detoxification of herbicides and toxic substances	Tobacco	Those overexpressing *gst/gpx* showed stimulated seedling growth under chilling stress compared to the wild type	Roxas et al., 1997
coda (Choline oxidase A) from *Arthrobacter globiformis* (a soil bacterium)	Biosynthesis of glycine betaine	Rice	Showed high levels of glycine betaine and those expressing the gene in the cytosol showed increased tolerance to photo inhibition under low temperature stress	Sakamoto and Murata, 1998
SCOF1 (Cold-inducible zinc finger protein) from soybean	Positive regulator of *SGBF-1*, which is a transcription factor, and *COR* (cold-regulated) gene	Tobacco	Those overexpressing *SCOF1* exhibited constitutive *COR* gene expression and significantly faster recovery after cold exposure in terms of whole plant morphology, fresh weight increase, and chlorophyll content	Kim et al., 2001
CuCOR19 (Citrus dehydrin)	Inhibition of lipid peroxidation by acting as a radical-scavenging protein	Tobacco	Increased cold tolerance as exhibited by earlier germination and better seedling growth as well as lesser malondialdehyde (MDA) production as compared to the control	Hara et al., 2003
DREB1A or *CBF3/rd29A* (Dehydration responsive element binding protein under the control of a stress-inducible promoter) from *A. thaliana*	*DREB1A* functions as a transcriptional activator in the expression of more than 40 stress-inducible genes that contain the C-repeat/dehydration responsive element in their promoters. *rd29A* is a promoter that initiates gene expression only in the presence of the stressor	Tobacco	Overexpression of *DREB1A* was induced by low temperature and resulted to higher stress tolerance; the *rd29A* promoter minimized the negative effects on plant growth due to constitutive overexpression	Kasuga et al., 2004

OSISAP1 (Zinc-finger protein) from rice	Stress-inducible gene	Tobacco	Seedlings overexpressing the transgene showed higher fresh weight under low temperature condition, higher gain in fresh weight during the recovery period, and healthier appearance compared to the wild type	Mukhopadhyay et al., 2004
FAD7/cor15a (Chloroplast-specific omega-3-fatty acid desaturase under the control of a cold-inducible promoter) from *A. thaliana*	*FAD7* gene product is for the desaturation of omega-3-fatty acid. *cor15a* is a promoter that initiates gene expression under cold conditions	Tobacco	Major trienoic fatty acids remained stable, chloroplast membrane ultrastructure was unaffected, high survival rate under prolonged cold stress exposure as opposed to the wild type	Khodakovskaya et al., 2006
OsMYB3R-2	Transcription factor targeting the cyclin gene *OsCycB1;1*, which is involved in the G2/M phase transition in the cell cycle at low temperature	Rice	Overexpressing transgenics showed higher percentage survival and normal growth during the recovery period after chilling treatment as well as higher survival rate after an extended exposure to low temperature	Ma et al., 2009
MYBS3	Transcription factor that activates expression of genes involved in various abiotic stress responses and/or tolerance; drives a novel cold signaling pathway	Rice	Overexpressing transgenics showed normal leaf phenotype and higher survival rate under cold stress	Su et al., 2010

CBF1 (CRT/ DRE binding factor 1)or *DREB1b* from *A. thaliana* and cotton	Transcriptional activators that control the expression of genes containing the C-repeat/ dehydration responsive element in their promoters; protection of photosystem II (PSII) and photosystem I (PSI) during low temperature stress at low irradiance	Tomato and tobacco	Overexpressing transgenic tomato had higher photochemical efficiency of photosystem II, oxidizable P700, superoxide dismutase activity, and non-photochemical quenching as well as lower MDA levels than the non-transgenics. Transgenic tobacco showed better membrane stability and higher free proline and soluble sugar levels at low temperature than the wild type	Zhang et al., 2011; Guo et al., 2011

Source: Sanghera, Wani, Hussain, and Singh (2011).

Table 3.1 shows that a number of genes involved in producing transgenic crops that are cold stress tolerant are sourced from *A. thaliana*. On the other hand, several genetic engineering experiments also used *A. thaliana* as the target host plant producing transgenic lines that either overexpress or under express certain genes in order to gain more insight into the mechanism of cold stress tolerance. Table 3.2 describes some of these genes.

Table 3.2: *A Selection of Genes Related to Cold Stress Tolerance Used to Produce Transgenic Arabidopsis Thaliana*

Gene/s (Gene Product)	**Function**	**Performance of Transgenic Plants**	**References**
COR15a (Cold-regulated gene)	Promotes freezing tolerance possibly by acting like a cryoprotective protein that protects membranes against freeze-induced damage	Constitutive expression reduced freeze-induced damage to photosystem II in chloroplasts of non-acclimated plants; enhanced the survival of protoplasts under freezing conditions *in vitro*	Artus et al., 1996
Antisense *AtProDH* (Proline dehydrogenase)	The antisense cDNA suppresses the enzyme, which catalyzes proline degradation	Maintained the levels of free proline and showed higher survival rate after freezing stress treatment	Nanjo et al., 1999
ALA1 (Aminophospholipid ATPase 1)	Maintains membrane symmetry, a putative aminophospholipid translocase	Antisense transgenics exhibited downregulation of *ALA1* and impaired growth under chilling temperatures compared to the wild type	Gomès et al., 2000

Osmyb4 from rice	Transcription factor that activates cold-inducible promoters such as *PAL2*, *ScD9*, *SAD*, and *COR15a*	Overexpressing transgenics showed higher survival rate and maintained membrane stability under freezing conditions; exhibited increased PSII stability despite photoinhibition under chilling stress	Vannini et al., 2003
ZAT12	Transcription factor that regulates the expression of 24 cold-responsive genes, nine of which were cold-induced and 15 were cold-repressed	Overexpressing non-acclimated transgenics showed improved survival under freezing treatment although the plans were smaller than the wild type	Vogel et al., 2004
ABI3 (Abscisic acid-insensitive protein 3)	Seed-specific transcriptional activator	Ectopic expression in vegetative tissues enhanced the ABA-induced activation of several cold and drought-responsive genes such as *RAB18LT129*, *LT130*, and *LT178*; improved ABA-induced freezing tolerance as indicated by membrane stability	Tamminen et al., 2008
ACBP6 (Acyl-coenzyme A binding protein 6)	Binds oleoyl-CoA and protects it from degradation by microsomal acyl hydrolases	Overexpressing transgenic maintained membrane stability and survived under freezing temperature; the seedlings also survived freezing treatment	Chen et al., 2008
AtCSP3 (Cold shock domain protein 3)	RNA chaperone that is induced by cold stress	Overexpressing transgenics displayed higher survival rate and more stable membranes at freezing temperature	Kim et al., 2009
Thp1 (Thermal hysteresis protein 1) from *Choristoneura fumiferana* (spruce budworm)	Antifreeze protein that decreases the freezing point of water below the melting point	Overexpressing transgenics exhibited better growth and membrane stability, lower MDA level, and higher proline level than wild type	Zhu et al., 2010
OsICE1 and *OsICE2* (Inducer of CBF expression) from rice	Transcription factors that directly bind and activate the expression of CBF/DREB1	Overexpression lead to higher survival rate sat freezing condition and up-regulation of cold-response genes under cold stress	Deng et al., 2017

CsINV5 (Vacuolar invertase 5) from tea plant	Regulates plant growth and development and responds to abiotic stresses such as drought and cold	Overexpression resulted to better membrane stability and photosystem fluorescence; higher levels of total soluble sugars, glucose, and fructose; and increased transcription of certain cold-related, osmotic stress-related, and water deprivation-related genes under cold conditions than the wild type	Qian et al., 2018

Source: Sanghera, Wani, Hussain, and Singh (2011).

As previously mentioned, certain genes that were previously described are involved in providing tolerance to more than one type of stress. These include genes for TFs, ROS scavenging machinery, compatible solute synthesis, heat shock proteins, LEA proteins or ion transport proteins. Table 3.3 lists some of these genes.

Table 3.3: *Genes That Are Involved in Conferring or Enhancing Tolerance to Multiple Stresses*

Gene (Gene Product)	**Stress Tolerance Provided**
OsEREBP1 (transcription factor)	Overexpression confers bacterial disease resistance and tolerance to drought and submergence
OsMYB3R-2 (transcription factor)	Overexpression increases tolerance to drought, cold, and salinity
Antisense *AtProDH* (Proline dehydrogenase)	Suppression of the enzyme improves tolerance to freezing and salinity
DREB1A (Dehydration-responsive element binding protein)	Drought, cold, and salinity
sod (Superoxide dismutase)	Drought and cold
OSISAP1 (Zinc finger protein)	Drought, cold, and salinity
sacB (Levan sucrase)	Drought and cold
GPX (Glutathione peroxidase)	Cold and salinity
GST (Glutathione S-transferase)	Drought, cold, and salinity
codA (Choline oxidase A)	Cold and salinity

3.2.2. Genetic Modification to Improve Yield

The development and cultivation of transgenic crops that show tolerance to both abiotic and biotic stresses reduce the damage caused by these stresses thus indirectly increasing the final yield when compared with conventional crops. Crops can also be genetically modified specifically for higher yield.

3.2.2.1. Enhancing plant growth

A transgenic eucalyptus containing a gene from *A. thaliana* showed improved growth. This encodes for a protein that accelerates plant growth by facilitating cell wall expansion. Therefore, the transgenic eucalyptus produces about 20% more wood than non-transgenic trees. Its time of maturity is also reduced, and it is ready for harvest in five and a half years instead of the usual seven. This is commercially available in Brazil since 2015 (Ledford, 2014; Norero, 2018).

3.2.2.2. Making photosynthesis more efficient

Photosynthesis is responsible for producing plant sugars, which are essential in the plant's growth and development. Therefore, improving photosynthetic efficiency is expected to increase plant yield whether the desired product in a crop is its fruit, seed or grain, wood or even roots. Several approaches have been attempted to achieve this end.

3.2.2.2.1. Switching From C3 to C4 Photosynthesis

An ongoing research on rice endeavors to make its photosynthesis more efficient by making it switch from C3 type of photosynthesis to C4, which makes maize and sugarcane more productive (see Figure 3.28). This entails genetically transforming rice to express the Kranz or wreath anatomy (see Figure 3.27) and the C4 metabolic enzymes, which are needed in order to maximize carbon fixation in the leaf and reduce photorespiration, which causes energy loss (see Figure 3.28). If this proves to be successful, rice yield is estimated to increase at approximately 30%–50% per hectare while using less water and fertilizer (IRRI, 2019; Bullis, 2015).

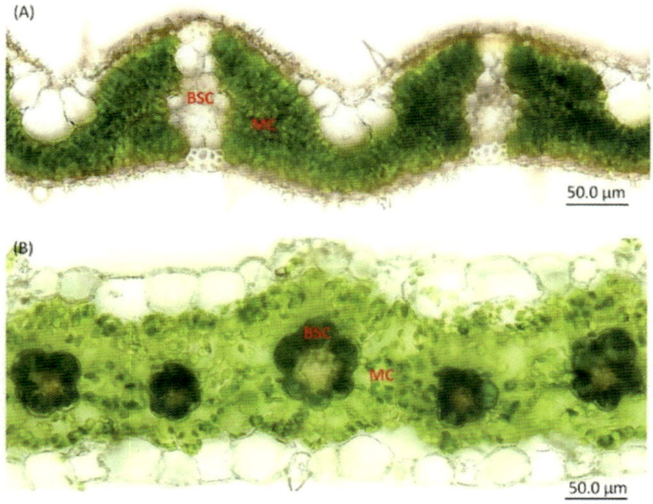

Figure 3.27: *(A) Rice is a C3 plant where more than 90% of the total chloroplasts are found in the mesophyll cells (MCs) and very few chloroplasts are in the bundle sheath cells (BSCs). (B) On the other hand, Setaria viridis is a C4 plant so it exhibits the Kranz or wreath anatomy where MCs surround the BSCs and equal quantity of chloroplasts are in both cell types.*

Source: Karki, Rizal, and Quick (2013).

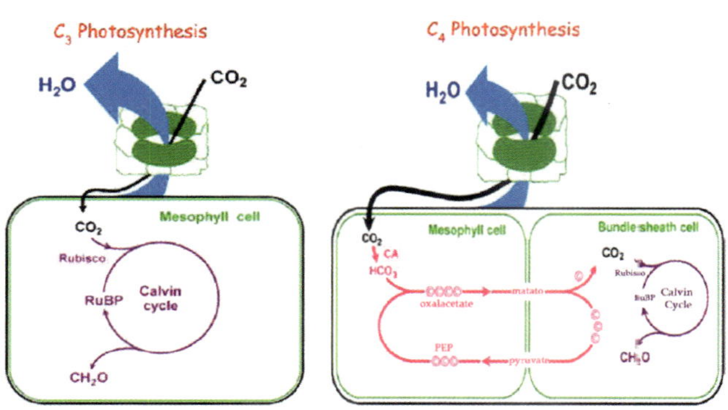

Figure 3.28: *In C3 photosynthesis, atmospheric $CO_2$is assimilated by Ribulose-1,5-bisphosphate carboxylase oxygenase (Rubisco) in the mesophyll cell to form a 3-carbon intermediate. This whole process is called the Calvin cycle. Rubisco can also fix O_2 when there is more of this gas than CO_2 in the intercellular spaces of the leaf tissue. This results to photorespiration which wastes energy produced by photosynthesis. In C4 photosynthesis, the mesophyll cell con-*

tains phosphoenolpyruvate carboxylase (PEPC) which fixes atmospheric CO_2 *generating a 4-carbon intermediate. This is decarboxylated to release* CO_2 *in the bundle sheath cell which contains Rubisco re-fixing it via the Calvin cycle. PEPC, unlike Rubisco, is insensitive to* O_2 *and so it concentrates* CO_2 *around Rubisco preventing photorespiration (Rizal et al., 2012).*

Source: Wang, Guo, Li, and Wang (2012).

Producing a C4 rice is an extremely complex and ambitious project. One step taken to convert the rice leaf anatomy to the C4 type is by increasing the number and size of chloroplasts in the bundle sheath cells (BSCs). This was attempted by introducing the *golden2-like* (*GLK*) gene from maize, which is needed for chloroplast development. This was then overexpressed in the mesophyll cells (MCs) by using the phosphoenolpyruvate carboxylase (PEPC) promoter of maize. It was also overexpressed in the BSCs by using the phosphoenolpyruvate carboxykinase (PCK) promoter of *Zoysia japonica* (zoysia grass). This resulted in increasing the functional chloroplast and mitochondria content of the BSCs (Karki et al., 2013; University of Oxford, 2017).

Another modification in the rice leaf anatomy that needs to be done is to reduce the space between the veins and increase the vein density from less than 6 veins per mm in C3 leaves to more than 7 veins per mm that is observed in C4 leaves. In addition, the Kranz anatomy in C4 leaves is characterized by the interconnection between mesophyll and BSCs via plasmodesmata. Some genes were identified to influence leaf development and possibly be manipulated to modify the rice leaf anatomy. These include *ACAULIS1* for leaf cell elongation, *CURLEY LEAF* for leaf curling, and *rotunda 1* for increasing free vein ending, opening venation pattern, and producing round leaves. Another gene is the *Scarecrow* in maize, the mutation of which resulted in increased number of BSCs, decreased minor veins, and reduced vein density. Mutation of the *Scarecrow* gene damaged the Kranz anatomy in C4 plants, but the same mutation produced normal C3 leaves. This indicates that regulation of this gene may be required for establishing Kranz anatomy (Karki et al., 2013; Slewinski et al., 2012).

The C3 and C4 leaves also differ metabolically (see Figure 3.28). C3 plants have a single carbon-fixing enzyme, Ribulose-1,5-bisphosphate

(RuBP) carboxylase oxygenase (Rubisco), that assimilates CO_2 via the Calvin cycle in the mesophyll cell. On the other hand, C4 plants have a different carbon-fixing enzyme in the mesophyll cell, phosphoenolpyruvate carboxylase (PEPC). Rubisco of C4 plants is in the bundle sheath cell where it is not exposed to O_2, thus preventing photorespiration. This means that creating C4 rice involves reducing Rubisco activity in the mesophyll cell and increasing or confining it in the BSCs. To accomplish this, genes that code for core C4 enzymes such as β carbonic anhydrase (CA), PEPC, pyruvate orthophosphate dikinase (PPDK), NADP-dependent malate dehydrogenase (NADP-MDH), and NADP-dependent malic enzyme (NADP-ME) were cloned from maize and introduced into rice to be overexpressed in the mesophyll cell cytosol (Karki et al., 2013).

Another candidate gene is for a transcription factor from maize, *Dof1*, which was introduced in wheat under the control of light-inducible Rubisco promoter. The transgenic wheat showed increased biomass and yield (Borisjuk et al., 2019). Moreover, genes for the transporters of metabolites between the mesophyll and the BSCs are also being transformed into rice. These include 2-oxoglutarate/malate transporter (OMT1), dicarboxylate transporter (DiT2), PEP/phosphate transporter (PPT1), mesophyll envelope protein (MEP), and triose-phosphate translocator (TPT) (Karki et al., 2013).

Photorespiration also needs to be reduced in order to make rice photosynthesis more efficient. When Rubisco catalyzes the assimilation of CO_2 with the 5-carbon RuBP, two molecules of 3-carbon compound, 3-phosphoglycerate (3-PGA), are formed. These proceed through the Calvin cycle to regenerate (RuBP) and to form a triose phosphate that is eventually used to produce sugar. However, when Rubisco reacts with O_2, RuBP is oxidized to form a molecule of 3-PGA and a molecule of a two-carbon compound, 2-phosphoglycolate. The 2-phosphoglycolate needs to be converted to 3-PGA through photorespiration. This involves a series of biochemical reactions occurring in the mitochondrion, peroxisome, and chloroplast. The downsides of photorespiration are the loss of previously fixed carbon and nitrogen and consumption of extra energy. One way to reduce photorespiration is by reducing the amount of glycine decarboxylase (GDC) in the mesophyll cell and restricting it in the bundle sheath cell. This ensures that glycine is decarboxylated only in the bundle sheath cell and thus, increasing CO_2 level in that cell. This was tested by introducing an artificial microRNA against the rice GDC-H subunit driven by PEPC promoter of maize in order to downregulate the *GDC* gene. Another approach was tested not in rice but in *A. thaliana*, which was transformed with genes for glycolate catabolic pathway from *Escherichia coli*. This biochemical pathway converts glycolate into glycerate. Five bacterial genes were introduced in a stepwise manner into *A. thaliana* and were specifically expressed in the chloroplast. These genes code for glycolate dehydrogenase, glyoxylate carboligase, and tartronic semialdehyde reductase. This reduced photorespiration and enhanced photosynthesis in *A. thaliana* (Kebeish et al., 2007; Karki et al., 2013)

A similar approach was tried on tobacco but in addition to genetically transforming the plant to perform the *E. coli* glycolate oxidation pathway, second, and third pathways were also engineered into the plant (see Figure 3.29). The second pathway used glycolate oxidase from *A. thaliana* and malate synthase from pumpkin and catalase from *E. coli* while the third one involved malate synthase from pumpkin and glycolate dehydrogenase from *Chlamydomonas reinhardtii* (green algae). The existing *PLGG1* chloroplast glycolate-glycerate transporter in tobacco was repressed by RNA interference (RNAi) in order to limit the native pathway (South et al., 2019).

Under greenhouse conditions, the first pathway, which is the *E. coli* pathway, increased biomass by approximately 13%. Transgenics expressing the second pathway did not show any improvement over the wild type. Pathway 3 altered photorespiration and increased photosynthetic rates as well as increased biomass by 18% without RNAi and 24% with RNAi. The transgenics were also planted on the field for two different growing seasons. Pathway 3 plants with RNAi had increased biomass by 20% compared to the wild type and in the following year, they exhibited 24% higher biomass. These plants also showed more efficient use of light during photosynthesis by 17%. The results for pathway 3 indicate that repressing *PLGG1* by RNAi prevents glycolate transport out of the chloroplast into the native photorespiratory pathway. This forces greater glycolate flux through the synthetic pathway in the chloroplast (South et al., 2019).

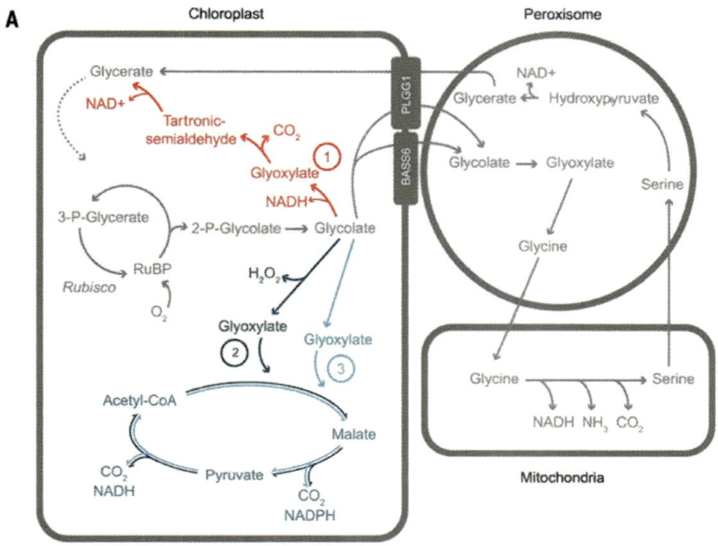

Figure 3.29: *Tobacco was genetically transformed to perform either one of three alternative photorespiratory pathways that metabolize glycolate in the chloroplast. Pathway 1 converts glycolate to glycerate using five genes from the Escherichia coli glycolate pathway that encode glycolate dehydrogenase, glyoxylate carboligase, and tartronic semialdehyde reductase. Pathway 2 and 3 converts glycolate to malate. Pathway 2 involved three transgenes encoding glycolate oxidase (from Arabidopsis thaliana), malate synthase (from pumpkin), and catalase (from E. coli). Catalase breaks down the hydrogen peroxide generated by glycolate oxidase. Finally, pathway 3 also contains pumpkin gene for malate synthase and Chlamydomonas reinhardtii glycolate dehydrogenase to avoid hydrogen peroxide generation when glycolate is converted to glyoxylate.*

Source: South et al. (2019).

3.2.2.2.2. Improving C3 Photosynthesis

Atmospheric CO_2 levels and air temperature are expected to increase by the year 2050. These changes are expected to limit C3 photosynthesis by decreasing the regeneration capacity of RuBP and the specificity of Rubisco for CO_2 relative to O_2 (Köhler et al., 2017).

A transgenic soybean was developed to adapt to this eventuality. It overexpresses the cyanobacterial (i.e., *Synechococcus*) gene for fructose-1,6/ sedoheptulose-1,7-biphosphatase (FBP/SBPase). SBPase catalyzes the dephosphorylation of sedoheptulose-1,7-biphosphate to sedoheptulose-7-phosphate at the branch point of RuBP regeneration and carbon export. Increased activity of this enzyme is expected to enhance RuBP regeneration of the transgenic plant (Köhler et al., 2017).

Wild type and overexpressing transgenic soybeans were grown under ambient and increased levels of CO_2 as well as under ambient and elevated temperatures. The transgenics exhibited significantly higher carbon assimilation than the wild type across all treatments. Under elevated concentration of $CO_2$and elevated temperature, the transgenics maintained their seed yield levels while the wild type had significantly reduced seed yield relative to the plants under increased CO_2 alone. These indicate that the transgenics had enhanced C3 photosynthesis and yield under expected future climate conditions that are expected to negatively affect food production (Köhler et al., 2017).

The same gene was used to genetically engineer Cadenza spring wheat. Transgenics grown in glasshouses showed higher yield by 15–20% (Le Page, 2016).

3.2.2.3. Enhancing Photoprotection

When light intensity gets too high, photoprotective mechanisms are activated to protect the photosynthetic antenna complexes from overexcitation. Excess excitation energy in the photosystem II (PSII) antenna complex can be dissipated as heat. This process is called nonphotochemical quenching (NPQ) of chlorophyll fluorescence. However, the rate of NPQ relaxation from quenched to unquenched state is slower than the rate of inducing the quenched state. In other words, it is faster to switch on the quenching mechanism than to switch it off. Thus, when light intensity changes from high to low such as when a cloud covers the sun or a leaf is shaded by another, the lack of light and the light wasted as heat via NPQ limit photosynthesis. It is estimated that these temporarily decreases CO_2 fixation by 7.5%–30% (Kromdijk et al., 2016; Devlin, 2016).

In order for NPQ to occur, the antennae have to change in conformation to the quenched state. This requires the presence of PSII subunit S (PsbS) and the occurrence of the xanthophyll cycle. The xanthophyll cycle involves the conversion of light-harvesting violaxanthin to the energy quenching antheraxanthin and zeaxanthin (see Figure 3.30). It was hypothesized that increasing PsbS and accelerating the xanthophyll cycle would result to faster decline in NPQ when there's sudden decrease in light intensity. This, in turn, enables maximizing CO_2assimilation efficiency by the plant when experiencing sudden changes in light intensity (Kromdijk et al., 2016).

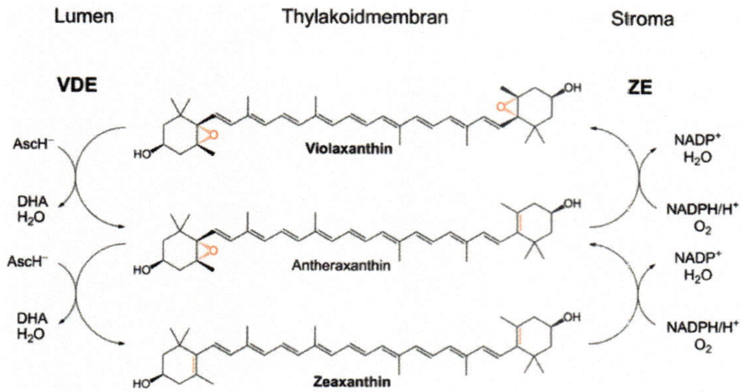

Figure 3.30: *During light stress, the xanthophyll cycle is induced which has a photoprotective role. This involves the removal of epoxy groups, as catalyzed by violaxanthin de-epoxidase (VDE), from violaxanthin producing the intermediate antheraxanthin, which is further de-epoxidized to form zeaxanthin. The reverse reaction which re-generates violaxanthin is catalyzed by zeaxanthin epoxidase (ZEP).*

Source: Yikrazuul, Public Domain.

Two genes from *A. thaliana* that code for xanthophyll cycle enzymes namely, *violaxanthin de-epoxidase* (*VDE*) and *zeaxanthin epoxidase* (*ZEP*) (see Figure 3.30), as well as the *PsbS* gene were all introduced to tobacco. The resulting transgenic tobacco exhibited faster relaxation of NPQ and recovery of CO_2 fixation rate when light intensity sharply decreased when compared to the wild type. Under field conditions, the transgenic tobacco plants showed higher total dry weight per plant as well as greater leaf area and plant height relative to the wild type (Kromdijk et al., 2016).

3.2.2.4. Improving Nitrogen Metabolism

Nitrogen is a significant plant nutrient that is essential in the biosynthesis of macromolecules like proteins and nucleic acids. Hence, nitrogen metabolism is expected to influence crop productivity.

While nitrogen gas (N_2) makes up the greatest proportion of the earth's atmosphere, it cannot be directly utilized by plants. Legumes have rhizobia in their root nodules that can perform N_2 fixation and initiates the conversion of this gas to ammonia. Ammonia, in turn, is converted by nitrifying microorganisms to nitrites or nitrates, which can be absorbed by the roots from the soil and then utilized by the plant for its biosynthetic processes. This process is called the nitrogen cycle (see Figure 3.31).

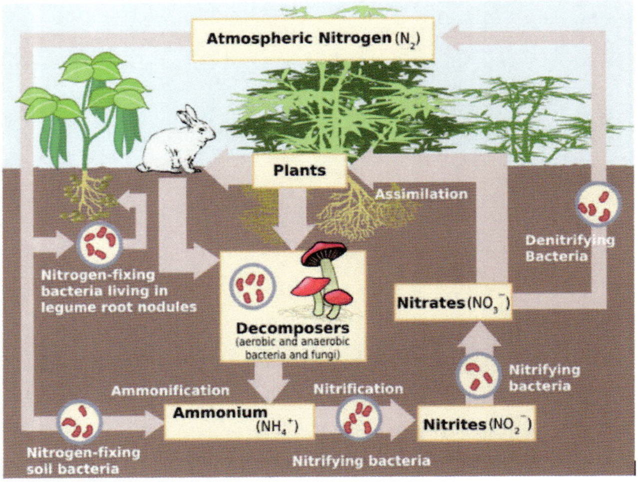

Figure 3.31: *The nitrogen cycle converts nitrogen gas (N_2) into a form that can be assimilated by plants. This involves a nitrogen-fixing reaction performed by symbiotic bacteria in the root nodules of legumes. The resulting ammonium is converted to nitrites then nitrates by nitrifying bacteria. The nitrates are then taken up by the plants to be used as nitrogen source for certain biosynthetic processes. The nitrates can also be converted into N_2 by denitrifying bacteria. When the plant or certain plant parts die and decompose, the nitrogen-containing compounds in the plants are broken down into ammonia.*

Source: Johann Dréo, Burkhard, Environmental Protection Agency, Raeky, Creative Commons License.

In soybean, ammonia is reduced into ureides such as allantoin and allantoic acid, which are exported from the nodules via the ureide permeases (UPS1) into the xylem. These are then transported via the xylem to supply nitrogen for the shoot. A common bean *UPS1* was used to genetically transform soybean and this was expressed in the cortex and endodermis cells of the transgenic soybean's nodules. These were grown in the greenhouse and it was observed that their rate of nitrogen delivery from nodules to shoot was significantly increased, growth rate was higher, and seed yield was greater by up to 36% relative to the wild type. Moreover, the transgenics exhibited higher levels of nodules, N_2 fixation per nodule, nitrogen assimilation in the nodule, and ureide synthesis. These results indicate that increasing ureide export from nodules is significant for N_2 fixation and nodule function. Therefore, improving it may lead to enhanced nitrogen nourishment for the plant and better shoot and seed development (Carter and Tegeder, 2016; Ferguson, 2016).

In wheat, *TaNAC2-5A* is a transcription factor that functions in nitrogen signaling. This was overexpressed in transgenic wheats leading to improved root growth and nitrate influx rate. As a result, the roots had greater capacity to uptake nitrate from the soil. Under conditions of severe nitrogen deficiency, the transgenics can use nitrogen more efficiently to produce shoot dry matter when compared to the wild type. Under field conditions, higher nitrogen levels in the shoot as well as higher grain yield were also observed in the transgenic plants relative to the wild type (He et al., 2015; Borisjuk, 2019).

Another transgenic wheat that was developed expressed *TaGS2-2Ab*. This gene codes for a major glutamine synthetase 2 (GS2) isozyme in the plastids of wheat leaf tissues during vegetative growth. GS catalyzes the first step of assimilating inorganic nitrogen into organic compounds. Transgenic expression of *TaGS2-2Ab* increased nitrogen acquisition by the roots, nitrogen uptake before and after flowering, and remobilization of nitrogen to grains possibly by upregulating the expression of nitrate transporters. During grain filling, the transgenic lines had increased leaf size and functional duration (i.e., delayed leaf senescence and extended photosynthesis), which may contribute to higher grain yield. When the generated transgenics were planted in the field, they showed higher grain yield, grain number per spike, and 1000-grain weight than the wild type in both low and high nitrogen conditions (Hu et al., 2018; Borisjuk, 2019).

Another benefit of developing transgenic plants that have a better ability to fix N_2, uptake, and assimilate nitrate or allocate nitrogen more efficiently throughout the different plant organs is that it can eliminate the need for nitrogen fertilizers. This can lessen the expense of farmers without compromising the quantity of their harvest.

Plants take up nitrogen either in the form of ammonium or nitrate or both. The ratio of these two chemical forms of nitrogen supply also influences cellular pH. In the case of rice, which is traditionally cultivated in flooded anaerobic soil conditions, ammonium is its main nitrogen source. Draining the soil may lead to ammonium-nitrate shift which can disrupt pH homeostasis. In rice, a nitrate transporter, OsNRT2.3, is involved in pH buffering, particularly its OsNRT2.3b splice isoform. This is expressed in the phloem and located in the plasma membrane. It has a regulatory motif on the cytosolic side that activates or inhibits nitrate transport by a pH-sensing mechanism. Transgenic rice that overexpressed *OsNRT2.3b* had improved pH-buffering capacity increasing the uptake of nitrogen, iron, and potassium. In field trials, the transgenic lines showed higher grain yield and nitrogen use efficiency by 40%. These results indicate that pH sensing by OsNRT2.3b has an important role in the plant's adaptation to varied forms of nitrogen supply and thus, enhances efficiency of nitrogen use as well as plant productivity (Fan et al., 2016).

3.2.2.5. Accelerating hybrid variety development

Hybrid varieties exhibit better tolerance to stresses and produce greater yield than their high yielding, resistant pure-bred parents. This is called hybrid vigor. Producing a hybrid variety through conventional breeding method takes about seven to 12 years depending on the crop (Bruins, 2016). Using genetic engineering can significantly shorten the time frame of producing a hybrid variety. This is the case in producing a hybrid variety of mustard in India called Dhara Mustard Hybrid-11 (DMH-11). This was genetically transformed with the *barnase*, *barstar*, and *bar* genes from *Bacillus amyloliquefaciens*. The *barnase* gene confers pollen sterility, which makes it easier to cross-pollinate the parental varieties. Meanwhile, the *barnstar* gene restores fertility in order to produce fertile hybrid plants and seeds. The *bar* gene codes for PAT that confers tolerance to glufosinate herbicides. The transgenic DMH-11 hybrid with these three genes has about 30% more yield than the traditional variety (Kumar and Jain, 2017).

3.2.3. Genetic Modification to Enhance Nutritional Content

Plants are major source of food for man. These contain nutrients that contribute to human health. However, a diet made up predominantly of a single crop would hardly provide adequate levels of nutrients. Thus, the ideal diet consists of a variety of food, but this is not accessible for the poor in developing countries. For this reason, scientists are attempting to fortify food crops to contain essential nutrients such as vitamins and minerals as well as macronutrients.

3.2.3.1. Increasing the micronutrient levels

Several cases of genetic transformation to improve the nutritional content of food crops involve increasing their vitamin or mineral content. While these are needed in relatively small amounts, millions suffer the health consequences of micronutrient deficiency.

3.2.3.1.1. Vitamin A

A popular example of a crop that is biofortified via genetic engineering is the Golden Rice. This was developed by Ingo Potrykus and Peter Beyer to have high β-carotene content giving it a golden yellow color (see Figure 3.32). β-carotene is one of the fat-soluble organic compounds that are collectively called Vitamin A. The rice endosperm can synthesize geranylgeranyl diphosphate, an early precursor of β-carotene but not phytoene, which is the first carotenoid precursor. The Golden Rice was engineered to have two genes from daffodil and one gene from *Erwinia* bacterium coding for enzymes in the carotenoid biosynthetic pathway (see Figure 3.33). The gene for phytoene synthase is from daffodil and it is under the control of a rice promoter to ensure that it is expressed in the rice endosperm only. This enzyme converts geranylgeranyl diphosphate into phytoene. The second gene is from *Erwinia*. It codes for phytoene desaturase, which converts phytoene to lycopene via a two-step process. Finally, another daffodil gene is for lycopene cyclase, which catalyzes the final step of converting lycopene to β-carotene in the endosperm (Toenniessen, 2002).

Figure 3.32: *The Golden Rice is genetically transformed in order to produce a significantly higher quantity of β-carotene in its endosperm. As a result, it has a golden yellow color as opposed to traditional rice.*

Source: International Rice Research Institute, Creative Commons License.

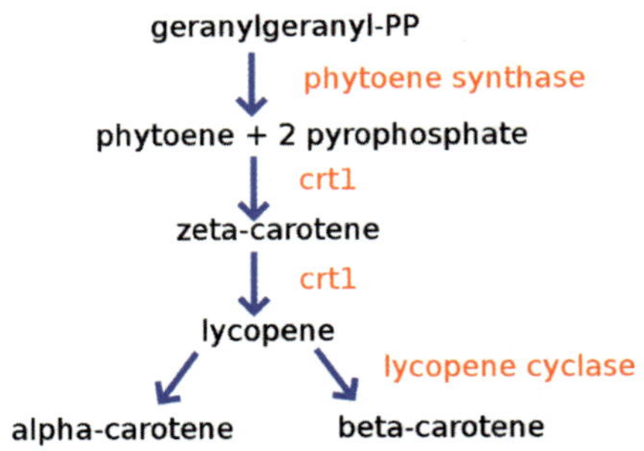

Figure 3.33: *The synthesis of β-carotene in the Golden Rice endosperm involves three enzymes. The first one is phytoene synthase which synthesizes phytoene from geranylgeranyl diphosphate. Phytoene desaturase (crtI) catalyzes the next two reactions which produce lycopene. Lycopene cyclase catalyzes the final step of converting lycopene to β-carotene.*

Source: Petaholmes and Gallet, Creative Commons License.

Eventually, Golden Rice 2 was developed, which contains as much as 37 g total carotenoids per gram of grain dry weight. This is much higher than the 1.6 g total carotenoid content of the first version of Golden Rice. Golden Rice 2 has the phytoene synthase gene from maize and the carotene desaturase gene from *Erwinia* or from another bacterium *Pantoea ananatis*. These transgenes are driven by the rice glutelin promoter to target their expression in the endosperm (ILSI International Food Biotechnology Committee, 2008; Health Canada, 2018). A similar approach was employed in fortifying canola with β-carotene (Newell-McGloughlin, 2008).

Other food crops, which are staples in the diet of most of the poor in the world, were fortified with β-carotene. These include maize and cassava. Triple-vitamin fortified maize was developed. The kernels of this transgenic maize contained 169 times more β-carotene, six times more ascorbate (vitamin C), and two times more folate (vitamin B_9) than the amount present in conventionally bred crops (Hefferon, 2015).

3.2.3.1.2. Folate (Vitamin B_9)

Plant foods are the main source of tetrahydrofolate in the human diet. Tetrahydrofolate is a complex molecule that is made up of three molecular components, namely pteridine, *para*-amino benzoic acid (pABA), and glutamate (see Figure 3.34A). Pteridine is synthesized in the plant cytosol while pABA is synthesized in the chloroplast (see Figure 3.34B). These two precursors are then used to produce tetrahydrofolate in the mitochondria. Folates are then derived from tetrahydrofolate (DellaPenna, 2007).

Tomato plants overexpressing GTP cyclohydrase I (GTPCHI), the first enzyme of pteridine synthesis (see Figure 3.34B), showed up to 140-fold increase in pteridine levels in their fruit. This resulted to a two-fold increase in folate levels but pABA levels decreased. For this reason, another transgenic tomato line was developed, which overexpressed in the fruit the enzyme that catalyzes the first step of PABA synthesis, aminodeoxychorismate synthase (ADCS) (see Figure 3.34B). The fruits contained 19 times more pABA than the wild type. When these two transgenic lines, the pteridine-overproducing line and the pABA-overproducing one, were crossed, the vine-ripened fruit of their progeny contained as much as 25 times more folate than the controls. Unripe fruits and those ripened by ethylene-gassing, which is the usual method of ripening commercial tomatoes, had up to 15-fold higher folate levels. The folate level in a single serving of this transgenic tomato fruit can already provide the complete daily requirement for adults (Díaz de la Garza et al., 2007).

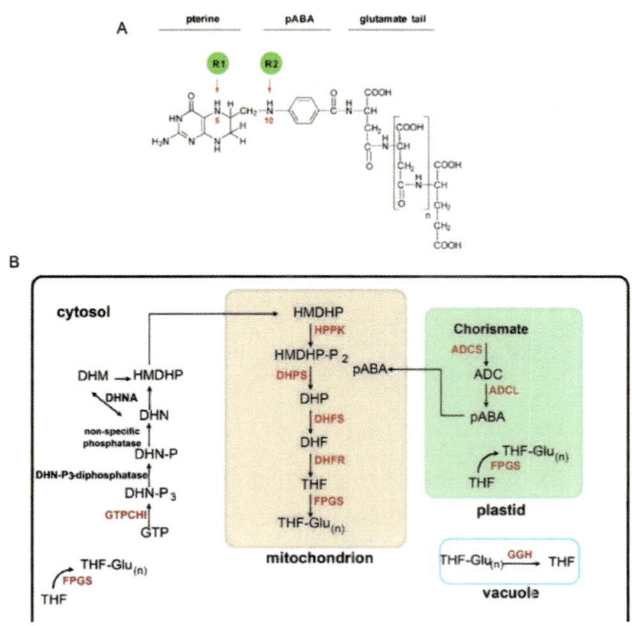

Figure 3.34: *(A) Folate or vitamin B_9 is a derivative of tetrahydrofolate, which is made up of pteridine, para-amino benzoic acid (pABA), and glutamate. (B) Pteridine is synthesized in the cytosol from GTP while pABA is made in the*

chloroplast (plastid). These two precursors are then transported into the mitochondrion where they are used to synthesize tetrahydrofolate (THF). The glutamate tail is then added to THF.

Source: Gorelova et al. (2019).

3.2.3.1.3. Iron

The most commonly deficient micronutrient is iron. Rice plants with the soybean gene for ferritin, an iron storage protein, was introduced into rice under the control of an endosperm-specific promoter. A gene for thermotolerant phytase from *Aspergillus fumigatus* was also introduced to improve iron absorption in the gastrointestinal tract. Phytase enzyme degrades phytic acid, which binds minerals forming an insoluble complex. The endogenous gene for cysteine-rich metallothionein-like protein was also overexpressed since cysteine enhances iron absorption. The resulting rice had grains with three times more iron than normal rice. In addition, they showed higher cysteine levels by as much as seven-fold and higher phytase levels by about 130 times (Lucca et al., 2002; Newell-McGloughlin, 2008).

Transgenic lettuce expressing ferritin was also developed. It contained from 1.2 to 1.7 higher levels of iron relative to the wild type. It also grew larger and faster during a 3-month period from germination (Goto et al., 2000).

3.2.3.2. Improving the amount of Macronutrients

Macronutrients are called such because they make up the bulk of a healthy diet. Certain food crops were genetically engineered to contain higher amounts of macronutrients.

3.2.3.2.1. Protein

The building blocks of proteins are amino acids. Most plants do not have the right balance of amino acids that are needed by humans. For instance, cereals like maize, wheat, and rice tend to be low in lysine. A transgenic maize with high levels of free lysine was developed by genetically transforming it with *dapA* gene from *Corynebacterium glutamicum*. This gene encodes a form of dihydropicolinate synthase that is insensitive to lysine feedback inhibition.

Another approach of producing food crops with increased protein levels is by introducing genes for proteins made up of elevated levels of the desired amino acid. A synthetic storage protein gene, *asp1*, coding for a protein rich in essential amino acids was introduced in sweet potato. The resulting plants showed a two-fold increase in total protein content in the leaves and a five-fold increase in the roots relative to the control plants. The amount of essential amino acids like methionine, threonine, tryptophan, isoleucine, and lysine significantly increased (Egnin and Prakash, 1997; Newell-McGloughlin, 2008).

Protein-enriched maize kernels were produced when transgenic maize was developed with modified floret development. There are two florets within each spikelet of the maize inflorescence. However, the lower floret is aborted early in its development, leaving the upper floret to mature. A transgenic line was developed with a bacterial gene for IPT, an enzyme that catalyzes cytokinin synthesis. The gene was expressed under the control of the *A. thaliana* promoter *SAG 12* (*senescence associated gene*) that is induced by senescence. The lower floret of the transgenic maize was rescued from being aborted and so there were two fertile florets in each spikelet. Both florets were fertilized resulting to two viable embryos, but the endosperms were fused to form just one kernel. The double embryo kernel has increased protein and oil content since these are mostly stored in the embryo (Young et al., 2004).

3.2.3.2.2. Carbohydrates

Photosynthesis in plants allows them to produce monosaccharides (e.g., fructose), disaccharides (e.g., sucrose), and polysaccharides (e.g., starch, fructan). Fructans can function as prebiotic agents improving colon health and reducing the incidence of colon cancer. The capacity to produce fructans was engineered into sugar beets by introducing the gene for 1-sucrose: sucrose fructosyl transferase (1-SST) from *Helianthus tuberosus* (Jerusalem artichoke). This enzyme catalyzes the first steps of converting sucrose into fructans. The taproot of the transgenic sugar beet had most of its stored sucrose converted into fructans (Sévenier et al., 1998; Newell-McGloughlin, 2008).

Fructans can form linear $\beta(2\rightarrow1)$polymers called inulins. Potatoes were transformed with the ability to manufacture inulin by the constitutive expression of the *1-SST* gene and fructan-fructan 1-fructosyltransferase (*1-FFT*) gene of *Cynara scolymus* (globe artichoke). The tubers of the transgenic potato plants were able to synthesize inulin molecules of all chain lengths. These made up 5% of the tuber dry weight (Hellwege et al., 2000).

3.2.3.2.3. Lipids

Medium chain fatty acids, with six to 10 carbon atoms, have been shown to minimize weight gain and fatty acid deposition while increasing the metabolic rate and physical endurance of test mice. However, these are minor components of most plant oil sources, like rapeseed (see Figure 3.35), with the exception of coconut and palm kernels. *Cuphea hookeriana* (a Mexican shrub) seeds contain high levels of medium chain fatty acids, specifically up to 75% of caprylate (with 8 carbon atoms) and caprate (with 10 carbon atoms). The *Ch FatB2*gene from this plant, which codes for acyl-ACP thioesterase, was overexpressed in the seed of transgenic canola. This enzyme has a role in determining chain length during fatty acid biosynthesis. The seeds of the transgenic canola exhibited a significant increase in the levels of the two medium chain fatty acids (Dehesh et al., 1996; Newell-McGloughlin, 2008).

Omega-3 fatty acids have several beneficial health-related functions such as, regulating inflammatory immune reactions, embryonic brain development, and development of cognitive function just after birth. These are associated with the prevention and treatment of heart disease, arthritis, autoimmune diseases, and cancer. These are traditionally sourced from fish oil and certain algae but not in canola oil. However, transgenic canola was developed, which contains this type of lipids. The transgenic plant expresses Δ12 and Δ6-fatty acid desaturases from *Mortierella alpina* (a commercially grown fungus) and Δ15-fatty acid desaturase from canola. The seeds accumulated up to 23% stearidonic acid (SDA), an omega-3 fatty acid, by weight. When the first transgenic generation was self-pollinated, the seeds of their offspring were evaluated for their fatty acid composition. SDA made up about 16% of the total fatty acids. The total omega-3 content (SDA and alpha-linolenic acid) constituted about 60% of the seed lipid content while omega-6 fatty acids made up approximately 22% (Ursin, 2003; Newell-McGloughlin, 2008).

Figure 3.35: *Canola is an edible variety of rapeseed that was developed by Canadian scientists. Its name is derived from Canada and "ola" (which means oil). Canola seeds (inset) are a major source of plant-based oil.*

Source: Carl Davies, CSIRO, Creative Commons License; Inset: USDA, Public Domain.

Omega-6 fatty acids have health benefits similar to their omega-3 counterparts. Gamma-linolenic acid (GLA) is an omega-6 fatty acid, which occurs in a limited number of plant species. Safflower (see Figure 3.36) is an oilseed crop containing high levels of linoleic acid (LA), another type of omega-6 fatty acid. It was genetically transformed with the Δ6-fatty acid desaturase gene from *Saprolegnia diclina* (a water mold). The Δ6-fatty acid desaturase enzyme catalyzes the synthesis of GLA from LA. The seed oil from the transgenic lines had GLA levels exceeding 77%. Another kind of transgenic safflower was developed by overexpressing Δ12- and Δ6-desaturases from the fungus *Mortierella alpine*. GLA levels in the resulting seed oil were around 55% (Nykiforuk et al., 2012; Newell-McGloughlin, 2008).

Figure 3.36: *Safflower is an oilseed crop with the most common varieties having yellow or orange flowers. Oil derived from its seeds (inset) contains a higher proportion of unsaturated fatty acids relative to saturated fatty acids.*

Source: Pseudoanas, Public Domain; Inset: Ahmad et al. (2012).

3.2.3.3. Increasing levels of therapeutic phytochemicals

A phytochemical broadly refers to a wide variety of compounds that occur naturally in plants. This may include the macronutrients and micronutrients, some examples of which have been previously discussed, but the term is usually used to pertain to plant-derived compounds with therapeutic properties such as anticarcinogenic, antimutagenic, anti-inflammatory, and antioxidant activities. Some plants have been engineered genetically to produce certain therapeutic metabolites.

One type of phytochemical is polyamine, which is an organic cation. It plays a role in regulating cell cycle and growth, thus influencing development as well as many physiological processes in the human body. Tomato naturally contains polyamines, but it could be genetically transformed to increase its polyamine levels. This was accomplished by introducing *Spe2*, a yeast S-adenosylmethionine decarboxylase gene, which was fused with a ripening-inducible promoter. The overexpression of the transgene resulted to increased levels of the polyamines spermidine and spermine in the ripe fruit. Lycopene levels were also higher in the transgenic fruit by 2.5 to 3 times than in the non-transgenic one (Mehta et al., 2002; Newell-McGloughlin, 2008).

Another example of a therapeutic phytochemical is resveratrol, a type of compound called stilbene. This compound has antioxidant, anti-cyclooxygenase, and anti-estrogenic activities. It also modulates lipid and lipoprotein metabolism as well as inhibits platelet aggregation. A diet rich in resveratrol may prevent atherosclerosis, carcinogenesis, and other degenerative diseases. Grapes, mulberries, and peanuts are rich in resveratrol as these have the stilbene synthase (StSy) enzyme, which is essential in stilbene synthesis. However, these are not usually found in a common person's diet unlike tomato, which is widely consumed. However, tomatoes do not have the *StSy* gene. They do, however, contain the precursors of resveratrol synthesis as these are the same substrates for flavonoid synthesis. Tomato was transformed with the *StSy* gene from grape under the transcriptional control of a constitutive promoter. Transgenic plants contained up to 53 μg/g FW of *trans*-resveratrol at the red stage of ripening. There was also an observed increase in the levels of the antioxidants ascorbate and glutathione, further enhancing the nutritional value of the transgenic fruit (Giovinazzo et al., 2004). Other plants such as alfalfa, wheat, and kiwi were also genetically transformed to produce resveratrol (Newell-McGloughlin, 2008).

Tomatoes are also engineered to express increased concentration of the antioxidant anthocyanin. Anthocyanin is thought to fight cardiovascular disease and to have anti-inflammatory properties. Blueberries and blackberries contain high amounts of anthocyanin, but these are seasonal and relatively expensive unlike tomatoes, which are more affordable. When two TFs from snapdragon were expressed in tomato, the resulting fruit contained anthocyanin at concentrations comparable to those found in blueberries and blackberries. When cancer-susceptible mice were fed with the transgenic tomato, they showed a significantly longer life span by up to 30% (Butelli et al., 2008; Hefferon, 2015). Similarly, when *SIMYB75*, a gene for a certain transcription factor in tomato, was overexpressed, the fruit of the transgenic tomato showed significantly higher levels of anthocyanin than the wild type. This resulted to the fruit becoming purple in color instead of the usual red (see Figure 3.37). Levels of phenolics, flavonoids, ethylene, and aroma volatiles such as aldehyde, phenylpropanoid-derived, and terpene volatiles were also substantially higher in the transgenic fruit. Analysis of the transcriptome revealed that SIMYB75 upregulated genes involved in ethylene signaling, phenylpropanoid and isoprenoid pathways (Jian et al., 2019).

Figure 3.37: *Overexpression of SIMYB75 in tomato significantly increased anthocyanin levels relative to the wild type. This resulted to purple fruits at different growth stages and smaller, purple seeds.*

Source: Jian et al. (2019).

3.2.4. Genetic Modification to Produce Recombinant Proteins

Plants, both food and non-food crops, have been engineered to produce therapeutic proteins such as monoclonal antibodies (mAbs), vaccine proteins, and anti-cancer agents. This field is also called molecular farming, which began from the need for safe and inexpensive biopharmaceuticals in developing countries. For instance, it is envisioned that food crops that produce vaccine proteins can be consumed directly or made into chips or powder juice and still elicit an immune response against the target pathogen. These are easily transportable and do not require refrigeration. In addition, this can be cultivated and require only partial processing unlike the typical fermentation systems needed for pharmaceutical manufacturing (Hefferon, 2015; Margolin et al., 2018).

Therapeutic proteins can be produced either by the integration and expression of the transgene in the nuclear genome of the plants or by transient expression of the transgene in plants. Developing transgenic plants require a long time and have resulted to low protein yield. On the other hand, transient expression can generate the proteins within 1 to 2 weeks. Plant viral vectors are usually used in this approach since they replicate efficiently resulting to high copy numbers of the transgene and their mRNA transcripts. Therefore, a substantial quantity of the corresponding protein is synthesized and accumulated (Lai et al., 2011). Another method employed is integrating the transgene in the chloroplast genome.

3.2.4.1. Transgenic expression of therapeutic proteins

Therapeutic proteins are produced by generating transgenic plants, which are envisioned to become edible vaccines. This is the method employed in producing vaccine against the parasite *Plasmodium vivax*, which causes malaria. A chimeric gene was developed containing merozoite surface protein-1 (MSP-1), an antigen considered to be a strong vaccine candidate, and circumsporozoite protein (CSP), which contains a B-cell epitope. This was expressed in *Brassica napus*. The chimeric recombinant protein exhibited 84.21% clinical sensitivity and 100% clinical specificity. The purified protein was administered to mice orally and it successfully induced antigen-specific IgG1 production. Moreover, it increased the production of the cytokines IL-12, TNF, and IFN-γ. These results indicate that the plant-derived protein has the potential as an antigen for diagnosis and for vaccine development against vivax malaria (Lee et al., 2011).

Carrots have also been genetically transformed to express *Mycobacterium tuberculosis* virulence proteins such as ESAT6 and CFP10. These proteins accumulated in the carrot storage roots. Oral administration of the recombinant antigens induced both cell-mediated (i.e., lymphocyte activation) and humoral (i.e., IgG production) immunities in the test mice. This transgenic carrot may serve as the platform for designing edible antituberculosis vaccine (Uvarova et al., 2013).

Another class of therapeutic proteins produced in transgenic plants is allergens, which can be used to induce tolerance to allergies such as bronchial asthma. The gene for a fragment of mite allergen called *Der p 1* was introduced into rice and the protein accumulated in the rice grain at approximately 90 μg/grain. When test mice were orally immunized with purified Der p 1, serum level of antigen-specific IgE significantly increased. When the test mice were fed the transgenic rice grain for seven days prior to immunization with recombinant Der p 1, the serum level of allergen-specific IgG was significantly lower than in the control mice. This means that tolerance to the allergen was successfully induced. Allergen-induced $CD4^+$T cell proliferation and production of Th2 cytokines *in vitro*, infiltration of eosinophils, neutrophils, and mononuclear cells into the airways and bronchial hyper responsiveness were also inhibited by oral vaccination (Suzuki et al., 2011).

3.2.4.2. Transient expression using plant viral vectors

Deconstructed plant viral vectors are commonly used to achieve temporary expression of the transgene. When designing the vector, the viral genome is fused with a plant promoter and viral coat protein genes are deleted to eliminate the risk of producing infectious virions. *A. tumefaciens* is commonly used to efficiently deliver the vector into the target plant cells without the entire plant being infected with the virus. Once inside the plant cells, the viral genome is transcribed and spliced to produce functionally infective replicon (Lai et al., 2011).

A geminiviral replicon vector based on the bean yellow dwarf virus was designed to contain genes for the Norwalk virus capsid protein (NVCP). A dual-replicon vector was also constructed containing genes for both the light chain and heavy chain of either humanized mAB against Ebola virus (6D8 mAb) or the West Nile virus (hE16 mAB). These were introduced into separate *A. tumefaciens* by electroporation. Then lettuce heads were infiltrated with these transgenic *A. tumefaciens*. NVCP was produced in the lettuce leaves with an average expression of approximately 0.2 mg/g leaf FW. This was extracted from the lettuce leaves and purified via low pH precipitation and anion exchange chromatography. This two-step purification process removed most of the host proteins, such as Rubisco, and enriched NVCP to greater than 95% purity. The purified NVCP self-assembled into empty virus-like particles (VLPs) making it a vaccine candidate (Lai et al., 2011).

The agroinfiltrated lettuce leaves also rapidly produced fully assembled 6D8 mAb and hE16 mAB accumulating up to 0.23–0.27 mg/g leaf FW. These proteins were extracted from the lettuce leaves and then purified via single affinity chromatography. Greater than 95% purity with intact light chain and heavy chain of both types of mABs were obtained. These mAB also proved to be functional as demonstrated by their capacity to specifically bind to their target viral protein (Lai et al., 2011).

3.2.4.3. Chloroplast-specific expression of transgenes

Integrating the gene of interest specifically into the chloroplast genome can ensure its containment so that transgenic plants can be planted in the field with reduced risk of contaminating wild types. This is because the chloroplast DNA is maternally inherited and is not involved in sexual reproduction. Chloroplasts are also very efficient bioreactors capable of producing 360 million doses of fully functional vaccine protein per acre of tobacco, for instance (Davoodi-Semiromi et al., 2010).

The chloroplast vectors with genes for cholera toxin-B subunit (CTB) of *Vibrio cholerae* fused with a gene for malaria vaccine antigen, either apical membrane antigen-1 (AMA1) or MSP1, were inserted into the chloroplast genomes of tobacco and lettuce by homologous recombination. Plants that successfully integrated the transgene in their chloroplast genome are described as transplastomic. Transplastomic lines of tobacco and lettuce successfully produced dimers, trimers, tetramers, and pentamers of CTB-AMA1 and CTB-MSP1 fusion proteins. A gram of mature leaf of transformed tobacco and lettuce contained up to 3.56 mg and 1.21 mg of CTB-AMA1 fusion protein, respectively. In terms of CTB-MSP1, a gram of mature leaf of tobacco and lettuce had 2.27 mg and 0.83 mg of CTB-MSP1, respectively. The pentameric form of CTB binds to GM_1ganglioside receptor *in vivo*. Binding assay showed that the CTB-AMA1 and CTB-MSP1 proteins derived from the transplastomic lines were fully functional and successfully bound GM_1. This indicates that the extracted proteins from both plants assembled properly into pentameric form. However, tobacco presents a disadvantage for human oral delivery studies since it contains nicotine or other alkaloids. On the other hand, lettuce leaves can be consumed raw (Davoodi-Semiromi et al., 2010).

Test mice were immunized either subcutaneously or orally with purified antigens or transplastomic tobacco leaves. Antibody levels were significant enough to completely inhibit the proliferation of malaria parasite. Mice that were challenged with cholera toxin were also protected 100% after oral immunization and 89% after subcutaneous immunization. Protection conferred by the immunization correlated with the intestinal and serum levels of CTB-specific immunoglobulin A (IgA) and IgG1 after oral vaccination and IgG1 levels after subcutaneous vaccination (Davoodi-Semiromi et al., 2010).

Lettuce and tobacco transcriptomic lines expressing cholera toxin B subunit-proinsulin (CTB-Pins) fusion proteins were also generated. These proteins accumulated up to ~16% and ~2.5% of total soluble proteins in tobacco and lettuce, respectively. Powdered leaves from the transcriptomic tobacco were administered orally to non-obese diabetic mice. The pancreas of treated mice showed decreased insulitis (i.e., lymphocyte infiltration) and significantly preserved β-cells that produce the insulin. These mice also exhibited lower glucose levels in their blood and urine as well as elevated serum levels of IgG1. Additionally, the pancreas of CTB-Pins-treated mice exhibited increased expression of immunosuppressive cytokines such as interleukin-4 (IL-4) and IL-10 (Ruhlman et al., 2007).

3.2.4.4. Plant-derived viral glycoproteins

The induction of neutralizing antibodies is the most consistent biomarker of vaccine efficacy. These antibodies target the glycoprotein spikes on the surface of enveloped viruses. Therefore, the development of glycoprotein-based antigens is highly desirable when producing vaccines against viruses. There are already several cases of successfully expressing virus glycoprotein in plants (see Figure 3.38). A number of strategies have also been developed to improve the quantity of these recombinant proteins obtained from plants. These include targeting the proteins to the secretory pathway, N-glycosylation, proteolytic maturation along the secretory pathway, and involving chaperones in the endoplasmic reticulum (Margolin et al., 2018).

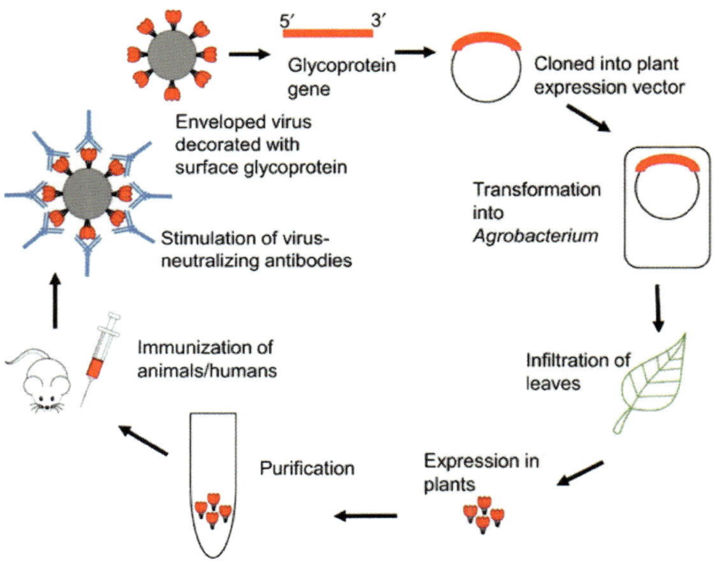

Figure 3.38. *Production of viral glycoproteins as vaccine immunogens involves introducing the glycoprotein gene of the virus into an expression vector. The target plant is then genetically transformed (Agrobacterium-mediated in this case) with this recombinant vector. The glycoproteins produced in the transgenic plant are extracted, purified from contaminating host proteins, and then used for immunization to elicit neutralizing antibodies against the virus.*

Source: Margolin et al. (2018).

The expression of viral glycoproteins in plants could either be transient, transplastomic, or transgenic. Table 3.4 presents some of these that were already tested in animals.

Table 3.4: *Viral Glycoproteins Produced in Plants*

Virus (Disease Caused)	Glyco-Protein	Expression Host	Expression Method and Yield	Antigen Tested in Animals and Immunization Method	Immunogenicity	References
Corona virus (Severe acute respiratory syndrome)	Spike protein (S1)	Tobacco root	Transgenic, not quantitated	Lyophilized tobacco root, parenteral.	IgG (after three immunizations and a booster dose)	Pogrebnyak et al., 2005
		Tomato fruit	Transgenic, not quantitated	Lyophilized ripe fruit, oral	IgA	
Dengue virus	Dengue 2 envelope domain III (D2EIII)	*Nicotiana benthamiana* leaf	Transient, 5.6 mg/kg fresh weight (FW)	Purified D2EIII, intramuscular	IgG and neutralization of dengue type 2 virus	Saejung et al., 2007
Ebola virus	Ebola immune complex (EIC: glycoprotein 1 fused with heavy chain of IgG monoclonal antibody)	*Nicotiana benthamiana* leaf	Transient, 50 mg/kg FW	Purified EIC, subcutaneous	IgG	Phoolcharoen et al., 2011
Human immunodeficiency virus (AIDS)	Soluble HIV-1 subtype C envelope glycoprotein 140 (gp140)	*Nicotiana benthamiana* leaf	Transient, 5–6 mg/kg FW.	Purified gp140, intramuscular	Binding and neutralizing antibodies	Margolin et al., 2019
Human papillomavirus (Cervical cancer)	HPV16L1 capsid protein with *Escherichia coli* heat labile enterotoxin B subunit (LTB)	Tobacco leaves	Transgenic, 0.22–0.31% of total soluble cell protein	Purified HPV16L1, oral	IgG, IgA, IFNγ/IL-4 $CD4^+$	Hongli et al., 2013
Influenza virus	H5N1 and H1N1 hemagglutinins (HA) virus-like particles.	*Nicotiana benthamiana* leaf	Transient, 50mg/kg FW (for both)	Purified HAs, intramuscular	Haemagglutination inhibition titers	D'Aoust et al., 2008
Rabies virus	Surface protein G	Maize kernels	Transgenic, 25 mg/kg FW	Ground kernels, oral	Neutralizing antibody	Loza-Rubio et al., 2012

Zika virus	Envelope (E) glycoprotein (his-tagged)	*Nicotiana benthamiana*	Transient, 160 mg/kg FW	Purified E glycoprotein, subcutaneous	IgG, IFNγ, IL-4, IL-6, neutralizing antibodies	Yang et al., 2017

Source: Margolin et al. (2018).

REFERENCES

1. AgFax.com. (2010). *Do You Know the Difference Between Glufosinate or Glyphosate?* https://agfax.com/news/2010/03/glufosinate-glyphosate-weed-resistance-0303.htm (Accessed on 18 December 2019).
2. Agronom. *Creative Commons Attribution-Share Alike 3.0 Unported License.* https://commons.wikimedia.org/wiki/File:Symptoms_corncobs_destruction_caused_by_Ostrinia_nubilalis.JPG, Inset: Keith Weller, Public Domain, https://commons.wikimedia.org/wiki/File:Corn_borer.jpg (Accessed on 18 December 2019).
3. Ahmad, M., Teong, L. K., Zafar, M., Sultana, S., Sadia, H., & Khan, M. A., (2012). Prospects and potential of green fuel from some non-traditional seed oils used as biodiesel. IntechOpen, Creative commons attribution 3.0 license. In: Fang, Z., (ed.), *Biodiesel-Feedstocks, Production, and Applications*. https://www.intechopen.com/books/biodiesel-feedstocks-production-and-applications/prospects-and-potential-of-green-fuel-from-some-non-traditional-seed-oils-use-as-biodiesel (Accessed on 18 December 2019).
4. Anand, A., & Jones, T., (2018). Advancing agrobacterium-based crop transformation and genome modification technology for agricultural biotechnology. *Current Topics in Microbiology and Immunology*, 418. doi: 10.1007/82_2018_97.
5. Anand, A., Bass, S. H., Wu, E., Wang, N., McBride, K. E., Annaluru, N., Miller, M., Hua, M., & Jones, T. J., (2018). An improved ternary vector system for *Agrobacterium*-mediated rapid maize transformation. *Plant Molecular Biology*, *97*(1/2), 187–200. https://doi.org/10.1007/s11103-018-0732-y (Accessed on 18 December 2019).
6. Anderson, J. A., Ellsworth, P. C., Faria, J. C., Head, G. P., Owen, M., Pilcher, C. D., Shelton, A. M., & Meissle, M., (2019). Genetically engineered crops: Importance of diversified integrated pest management for agricultural sustainability. Creative commons attribution 4.0 international license. *Frontiers in Bioengineering and Biotechnology*, *7*, 24. doi: 10.3389/fbioe.2019.00024.
7. Artus, N. N., Uemura, M., Steponkus, P. L., Gilmour, S. J., Lin, C., & Thomashow, M. F., (1996). Constitutive expression of the cold-regulated *Arabidopsis thaliana COR15a* gene affects both chloroplast and protoplast freezing tolerance. *Proceedings of the National Academy of Sciences of the United States of America*, *93*(23), 13404–13409. doi:

10.1073/pnas.93.23.13404.

8. Badr, Y., Bahieldin, A., Aziz, M. A., Yehia, M. A., El-Magd, A. A., & Madkour, M. A., (2004). A modified protocol for laser-mediated gene transfer in wheat. *Arab Journal of Biotechnology*, *7*(2), 299–304.
9. Baltes, N. J., Gil-Humanes, J., & Voytas, D. F., (2017). Genome engineering and agriculture: Opportunities and challenges. In: Weeks, D. P., & Yang, B., (ed.), *Progress in Molecular Biology and Translational Science: Gene Editing in Plants* (Vol. 149, pp. 1–26). Academic Press: London, U. K. https://doi.org/10.1016/bs.pmbts.2017.03.011 (Accessed on 18 December 2019).
10. Bidney, D., Scelonge, C., Martich, J., Burrus, M., Sims, L., & Huffman, G., (1992). Microprojectile bombardment of plant tissues increases transformation frequency by *Agrobacterium tumefaciens*. *Plant Molecular Biology*, *18*(2), 301–313.
11. Blomme, G., Dita, M., Jacobsen, K. S., Vicente, L. P., Molina, A., Ocimati, W., Poussier, S., & Prior, P., (2017). Bacterial diseases of bananas and enset: Current state of knowledge and integrated approaches toward sustainable management. Creative commons attribution 4.0 international license. *Frontiers in Plant Science, 8*, 1290. doi: 10.3389/fpls.2017.01290.
12. Borisjuk, N., Kishchenko, O., Eliby, S., Schramm, C., Anderson, P., Jatayev, S., Kurishbayev, A., & Shavrukov, Y., (2019). Genetic modification for wheat improvement: From transgenesis to genome editing. *BioMed Research International*, 6216304. https://doi.org/10.1155/2019/6216304 (Accessed on 18 December 2019).
13. Bruins, M., (2016). Promising techniques to accelerate innovation in plant breeding. *New Breeding Techniques*, *3*(1). https://european-seed.com/2016/03/new-breeding-techniques/ (Accessed on 18 December 2019).
14. Bullis, K., (2015). Supercharged photosynthesis. *MIT Technology Review*. https://www.technologyreview.com/s/535011/supercharged-photosynthesis/ (Accessed on 18 December 2019).
15. Butelli, E., Titta, L., Giorgio, M., Mock, H. P., Matros, A., Peterek, S., Schiljen, E. G. W. M., Hall, R. D., Bovy, A. G., Luo, J., & Martin, C., (2008). Enrichment of tomato fruit with health-promoting anthocyanins by expression of select transcription factors. *Nature Biotechnology*, *26*, 1301–1308.

16. *C.M. Creative Commons Attribution-Share Alike 3.0 Unported License.* https://commons.wikimedia.org/wiki/File:Agrobacterium_tumefaciens_Forsythie.jpg (Accessed on 18 December 2019).
17. Carbonari, C. A., Latorre, D. O., Gomes, G. L. G. C., Velini, E. D., Owens, D. K., Pan, Z., & Dayan, F. E., (2016). Resistance to glufosinate is proportional to phosphinothricin acetyl transferase expression and activity in LibertyLink® and Widestrike® cotton. *Planta*, *243*, 925–933. doi: 10.1007/s00425-015-2457-3.
18. Carter, A. M., & Tegeder, M., (2016). Increasing nitrogen fixation and seed development in soybean requires complex adjustments of nodule nitrogen metabolism and partitioning processes. *Current Biology*, *26*(15), 2044–2051. https://doi.org/10.1016/j.cub.2016.06.003 (Accessed on 18 December 2019).
19. Carter, M., & Shieh, J., (2015). *Guide to Research Techniques in Neuroscience* (2nd edn.). Academic Press: London, U. K.
20. Chen, C. P., Chou, J. C., Liu, B. R., Chang, M., & Lee, H. J., (2007). Transfection and expression of plasmid DNA in plant cells by an arginine-rich intracellular delivery peptide without protoplast preparation. *FEBS Letters*, *581*(9), 1891–1897.
21. Chen, Q. F., Xiao, S., & Chye, M. L., (2008). Overexpression of the Arabidopsis 10-kilodalton acyl-coenzyme A-binding protein ACBP6 enhances freezing tolerance. *Plant Physiology*, *148*(1), 304–315. doi: 10.1104/pp.108.123331.
22. Chen, X., Lu, X., Shu, N., Wang, D., Wang, S., Wang, J., Guo, L., Guo, X., Fan, W., Lin, Z., & Ye, W., (2017). *GhSOS1*, a plasma membrane Na+/H+ antiporter gene from upland cotton, enhances salt tolerance in transgenic *Arabidopsis thaliana*. Creative commons attribution 4.0 international license. *PLoS One*, *12*(7), e0181450. https://doi.org/10.1371/journal.pone.0181450 (Accessed on 18 December 2019).
23. Ciencias Españolas and KoS. *Creative Commons Attribution-Share Alike 3.0 Unported License.* https://commons.wikimedia.org/wiki/File:Genetic_engineering_logo.png (Accessed on 18 December 2019).
24. Conrow, J., (2017). *GMO Banana Offers Hope for Disease and Pest Resistance*. Cornell Alliance for Science. https://allianceforscience.cornell.edu/blog/2017/04/gmo-banana-offers-hope-for-disease-and-pest-resistance/ (Accessed on 18 December 2019).
25. *Creative Commons Attribution Share-Alike 2.5 Canada License.*

https://home.cc.umanitoba.ca/~frist/PLNT2530/108/8b_DirectTransformation.pdf (Accessed on 18 December 2019).

26. CSIRO (Commonwealth Scientific and Industrial Research Organization), (2017). *Case Study: Insect Protected Cowpeas*. https://www.csiro.au/en/Research/AF/Areas/Food-security/Insect-protected-cowpeas (Accessed on 18 December 2019).
27. CSIRO. *Creative Commons Attribution 3.0 Unported License*. https://www.csiro.au/en/Research/AF/Areas/Food-security/Insect-protected-cowpeas (Accessed on 18 December 2019).
28. D'Aoust, M. A., Lavoie, P. O., Couture, M. M. J., Trépanier, S., Guay, J. M., Dargis, M., Mongrand, S., Landry, N., Ward, B. J., & Vézina, L. P., (2008). Influenza virus-like particles produced by transient expression in *Nicotiana benthamiana* induce a protective immune response against a lethal viral challenge in mice. *Plant Biotechnology Journal*, *6*(9), 930–940. https://doi.org/10.1111/j.1467-7652.2008.00384.x (Accessed on 18 December 2019).
29. Dale, J., James, A., Paul, J. Y., Khanna, H., Smith, M., Peraza-Echeverria, S., Garcia-Bastidas, F., Kema, G., Waterhouse, P., Mengersen, K., & Harding, R., (2017). Transgenic Cavendish bananas with resistance to Fusarum wilt tropical race 4. Creative commons attribution 4.0 international license. *Nature Communications*, *8*, 1496. https://doi.org/10.1038/s41467-017-01670-6 (Accessed on 18 December 2019).
30. Dalla, C. L., Malnoy, M., & Gribaudo, I., (2017). Breeding next generation tree fruits: Technical and legal challenges. Creative commons attribution 4.0 international license. *Horticulture Research, 4*(17067). doi: 10.1038/hortres.2017.67.
31. Darbani, B., Farajnia, S., Toorchi, M., Zakerbostanabad, S., Noeparvar, S., & Stewart, C. N. Jr., (2008). DNA-delivery methods to produce transgenic plants. *Biotechnology*, *7*, 385–402. doi: 10.3923/biotech.2008.385.402.
32. Davies, C. (2019) *CSIRO, Creative Commons Attribution 3.0 Unported License*. http://www.scienceimage.csiro.au/image/3974 (Accessed on 18 December 2019).
33. Davoodi-Semiromi, A., Schreiber, M., Nalapalli, S., Verma, D., Singh, N. D., Banks, R. K., Chakrabarti, D., & Daniell, H., (2010). Chloroplast-derived vaccine antigens confer dual immunity against cholera and malaria by oral or injectable delivery. *Plant Biotechnology Journal*, *8*(2), 223–242. https://doi.org/10.1111/j.1467-7652.2009.00479.x

(Accessed on 18 December 2019).

34. Dehesh, K., Jones, A., Knutzon, D. S., & Voelker, T. A., (1996). Production of high levels of 8:0 and 10:0 fatty acids in transgenic canola by overexpression of *Ch FatB2*, a thioesterase cDNA from *Cuphea hookeriana*. *The Plant Journal*, *9*(2), 167–172.
35. DellaPenna, D., (2007). Biofortification of plant-based food: Enhancing folate levels by metabolic engineering. *Proceedings of the National Academy of Sciences of the United States of America*, *104*(10), 3675–3676. doi: 10.1073/pnas.0700640104.
36. Deng, C., Ye, H., Fan, M., Pu, T., & Yan, J., (2017). The rice transcription factors *OsICE* confer enhanced cold tolerance in transgenic *Arabidopsis*. *Plant Signaling and Behavior*, *12*(5), e1316442. https://doi.org/10.1080/15592324.2017.1316442 (Accessed on 18 December 2019).
37. Devlin, H., (2016). Plants modified to boost photosynthesis produce greater yields, study shows. *The Guardian*. https://www.theguardian.com/science/2016/nov/17/plants-genetically-modified-to-boost-photosynthesis-produce-greater-yields-study-shows (Accessed on 18 December 2019).
38. Díaz de la Garza, R. I., Gregory, J. F. III, & Hanson, A. D., (2007). Folate biofortification of tomato fruit. *Proceedings of the National Academy of Sciences of the United States of America*, *104*(10), 4218–4222. doi: 10.1073/pnas.0700409104.
39. DiLeo, M. (2012) *Monsanto's GM Drought Tolerant Corn*. http://thescientistgardener.blogspot.com/2012/08/monsantos-gm-drought-tolerant-corn.html (Accessed on 18 December 2019).
40. Dréo, J.; Burkhard; Environmental Protection Agency; & Raeky . (2009). *Creative Commons Attribution, Share Alike 3.0 Unported License*. https://commons.wikimedia.org/wiki/File:Nitrogen_Cycle.svg (Accessed on 18 December 2019).
41. Egnin, M., & Prakash, C. S., (1997). Transgenic sweet potato expressing a synthetic storage protein gene exhibits high level of total protein and essential amino acids. *Vitro Cell Developmental Biology*, *33*, 52A.
42. Eyal, B. (2009) *Creative Commons Attribution-Share Alike 3.0 Unported License*. https://commons.wikimedia.org/wiki/File:Sonicator.jpg (Accessed on 18 December 2019).
43. Fan, X., Tang, Z., Tan, Y., Zhang, Y., Luo, B., Yang, M., Lian, X., Shen, Q., Miller, A. J., & Xu, G., (2016). Overexpression of a pH-sensitive

nitrate transporter in rice increases crop yields. *Proceedings of the National Academy of Sciences of the United States of America*, *113*(26), 7118–7123. https://doi.org/10.1073/pnas.1525184113 (Accessed on 18 December 2019).

44. Ferguson, W., (2016). Soybean nitrogen breakthrough could help feed the world. *Washington State University: WSU Insider*. https://news.wsu.edu/2016/10/10/soybean-nitrogen-breakthrough-help-feed-world/ (Accessed on 18 December 2019).
45. Gatehouse, A. M. R., Ferry, N., Edwards, M. G., & Bell, H. A., (2011). Insect-resistant biotech crops and their impacts on beneficial arthropods. *Philosophical Transactions of the Royal Society B: Biological Sciences*, *366*(1569), 1438–1452. doi: 10.1098/rstb.2010.0330.
46. Gatehouse, J. A., (2008). Biotechnological prospects for engineering insect-resistant plants. *Plant Physiology*, *146*(3), 881–887. doi: 10.1104/pp.107.111096.
47. Gelvin, S. B., (2003). *Agrobacterium*-mediated plant transformation: The biology behind the "gene-jockeying" tool. *Microbiology and Molecular Biology Reviews*, *67*(1), 16–37. doi: 10.1128/MMBR.67.1.16-37.2003.
48. Gelvin, S. B., (2005). Viral-mediated plant transformation gets a boost. *Nature Biotechnology*, *23*(6), 684–685. doi: 10.1038/nbt0605-684.
49. Giovinazzo, G., D'Amico, L., Paradiso, A., Bollini, R., Sparvoli, F., & DeGara, L., (2004). Antioxidant metabolite profiles in tomato fruit constitutively expressing the grapevine stilbene synthase gene. *Plant Biotechnology Journal*, *3*(1), 57–69. https://doi.org/10.1111/j.1467-7652.2004.00099.x (Accessed on 18 December 2019).
50. Gisbert, C., Rus, A. M., Bolarin, M. C., López-Coronado, J. M., Arrillaga, I., Montesinos, C., Caro, M., Serrano. R., & Moreno, V., (2000). The yeast *HAL1* gene improves salt tolerance of transgenic tomato. *Plant Physiology*, *123*(1), 393–402.
51. Gomès, E., Jakobsen, M. K., Axelsen, K. B., Geisler, M., & Palmgren, M. G., (2000). Chilling tolerance in Arabidopsis involves ALA1, a member of a new family of putative aminophospholipid translocases. *Plant Cell*, *12*(12), 2441–2453.
52. Gorelova, V., Bastien, O., De Clecrck, O., Lespinats, S., Rébeille, F., & Van Der Straeten, D., (2019). Evolution of folate biosynthesis and metabolism across algae and land plant lineages. Creative commons

attribution 4.0 international license. *Scientific Reports, 9*, 5731. https://doi.org/10.1038/s41598-019-42146-5 (Accessed on 18 December 2019).

53. Goto, F., Yoshihara, T., & Saiki, H., (2000). Iron accumulation and enhanced growth in transgenic lettuce plants expressing the iron-binding protein ferritin. *Theoretical and Applied Genetics, 100*(5), 658–664.
54. Green, J. M., & Owen, M. D. K., (2011). Herbicide-resistant crops: Utilities and limitations for herbicide-resistant weed management. *Journal of Agricultural and Food Chemistry, 59*(11), 5819–5829. doi: 10.1021/jf101286h.
55. Guo, H. M., Li, Z. C., Zhang, H., Xin, Y. Z., & Cheng, H. M., (2011). Cloning of cotton CBF gene for cold tolerance and its expression in transgenic tobacco. *Acta Agronomica Sinica, 37*(2), 286–293. https://doi.org/10.1016/S1875-2780(11)60009-6 (Accessed on 18 December 2019).
56. Gupta, A. S., Heinen, J. L., Holaday, A. S., & Allen, R. D., (1993). Increased resistance to oxidative stress in transgenic plants that overexpress chloroplastic Cu/Zn superoxide dismutase. *Proceedings of the National Academy of Sciences of the United States of America, 90*(4), 1629–1633. doi: 10.1073/pnas.90.4.1629.
57. Gupta, P. K., Balyan, H. S., & Gahlaut, V., (2017). QTL analysis for drought tolerance in wheat: Present status and future possibilities. Creative commons attribution 4.0 international license. *Agronomy, 7*(1), 5. https://doi.org/10.3390/agronomy7010005 (Accessed on 18 December 2019).
58. Gupta, R. C., & Crissman, J. W., (2013). Safety assessment including current and emerging issues in toxicologic pathology. Haschek, W. M., Rousseaux, C. G., & Wallig, M. A., (eds.), *Haschek and Rousseaux's Handbook of Toxicologic Pathology*. Academic Press: London, U. K.
59. Hagio, T., (1998). Optimizing the particle bombardment method for efficient genetic transformation. *JARQ, 32*, 239–247.
60. Hara, M., Terashima, S., Fukaya, T., & Kuboi, T., (2003). Enhancement of cold tolerance and inhibition of lipid peroxidation by citrus dehydrin in transgenic tobacco. *Planta, 217*(2), 290–298.
61. He, X., Qu, B., Li, W., Zhao, X., Teng, W., Ma, W., Ren, Y., Li, B., Li, Z., & Tong, Y., (2015). The nitrate-inducible NAC transcription factor

TaNAC2-5A controls nitrate response and increases wheat yield. *Plant Physiology*, *169*, 1991–2005. https://doi.org/10.1104/pp.15.00568 (Accessed on 18 December 2019).

62. Health Canada, (2018). *Novel Food Information-Provitamin A Biofortified Rice Event GR2E (Golden Rice)*. https://www.canada.ca/en/health-canada/services/food-nutrition/genetically-modified-foods-other-novel-foods/approved-products/novel-food-information-golden-rice-gr2e.html (Accessed on 18 December 2019).
63. Hefferon, K. L., (2015). Nutritionally enhanced food crops, progress and perspectives. *International Journal of Molecular Sciences*, *16*(2), 3895–3914. doi: 10.3390/ijms16023895.
64. Hellwege, E. M., Czapla, S., Jahnke, A., Willmitzer, L., & Heyer, A. G., (2000). Transgenic potato (*Solanum tuberosum*) tubers synthesize the full spectrum of inulin molecules naturally occurring in globe artichoke (*Cynara scolymus*) roots. *Proceedings of the National Academy of Sciences of the United States of America*, *97*(15), 8699–8704. doi: 10.1073/pnas.150043797.
65. Hongli, L., Xukui, L., Ting, L., Wensheng, L., Lusheng, S., & Jin, Z., (2013). Trangenic tobacco expressed HPV16-L1 and Lt-B combined immunization induces strong mucosal and systemic immune responses in mice. *Human Vaccines and Immunotherapeutics*, *9*(1), 83–89. doi: 10.4161/hv.22292.
66. Hu, M., Zhao, X., Liu, Q., Hong, X., Zhang, W., Zhang, Y., Sun, L., Li, H., & Tong, Y., (2018). Transgenic expression of plastidic glutamine synthetase increases nitrogen uptake and yield in wheat. *Plant Biotechnology Journal*, *16*(11), 1858–1867. https://doi.org/10.1111/pbi.12921 (Accessed on 18 December 2019).
67. Hwang, H. H., Yu, M., & Lai, E. M., (2017). *Agrobacterium*-mediated plant transformation: Biology and applications. *The Arabidopsis Book*, 15. https://doi.org/10.1199/tab.0186 (Accessed on 18 December 2019).
68. ILSI International Food Biotechnology Committee, (2008). Golden rice 2. *Comprehensive Reviews in Food Science and Food Safety*, *7*(1), 92–98. http://doi.wiley.com/10.1111/j.1541-4337.2007.00029_7.x (Accessed on 18 December 2019).
69. InTeGrate (Interdisciplinary Teaching about Earth for a Sustainable Future), (2018). *Herbicide Resistant Crops*. https://serc.carleton.edu/integrate/teaching_materials/food_supply/student_materials/1189 (Accessed on 18 December 2019).

70. International Rice Research Institute. *Creative Commons Attribution 2.0 Generic License.* https://www.flickr.com/photos/ricephotos/5516789000/in/set-72157626241604366 (Accessed on 18 December 2019).
71. IoOALP. *Creative Commons Attribution-Share Alike 4.0 International License.* https://commons.wikimedia.org/wiki/File:Excimer_laser_1.jpg (Accessed on 18 December 2019).
72. IRRI (International Rice Research Institute), (2019). *The C4 Rice Project.* http://photosynthome.irri.org/C4rice/index.php/19-about/57-science-of-c4-rice (Accessed on 18 December 2019).
73. ISAAA (International Service for the Acquisition of Agri-biotech Applications), (2008). *Biotechnology for the Development of Drought Tolerant Crops.* https://www.isaaa.org/resources/publications/pocketk/32/default.asp (Accessed on 18 December 2019).
74. ISAAA (International Service for the Acquisition of Agri-biotech Applications), (2018a). *Bt Eggplant.* https://www.isaaa.org/resources/publications/pocketk/48/default.asp (Accessed on 18 December 2019).
75. ISAAA (International Service for the Acquisition of Agri-biotech Applications), (2018b). *Herbicide Tolerance Technology: Glyphosate and Glufosinate.* https://isaaa.org/resources/publications/pocketk/10/default.asp (Accessed on 18 December 2019).
76. Isaac, N., (2018). *Nigeria's GMO Cowpea Wins Medical, Religious and Civic Support.* Cornell alliance for science. https://allianceforscience.cornell.edu/blog/2018/12/nigerias-gmo-cowpea-wins-medical-religious-civic-support/ (Accessed on 18 December 2019).
77. Ismail, A. M., (2018). Submergence tolerance in rice: Resolving a pervasive quandary. *New Phytologist, 218*(4), 1558–1569. https://doi.org/10.1111/nph.15188 (Accessed on 18 December 2019).
78. Iuchi, A., Kobayashi, M., Taji, T., Naramoto, M., Seki, M., Kato, T., Tabata, S., Kakubari, Y., Yamaguchi-Shinozaki, K., & Shinozaki, K., (2001). Regulation of drought tolerance by gene manipulation of 9-*cis*-epoxycarotenoid dioxygenase, a key enzyme in abscisic acid biosynthesis in *Arabidopsis*. *The Plant Journal, 27*(4), 325–333. https://doi.org/10.1046/j.1365-313x.2001.01096.x (Accessed on 18 December 2019).
79. Jian, W., Cao, H., Yuan, S., Liu, Y., Lu, J., Lu, W., Li, N., Wang, J., Zou, J., Tang, N., Xu, C., Cheng, Y., Gao, Y., Xi, W., Bouzayen, M., &

Li, Z., (2019). SIMYB75, an MYB-type transcription factor, promotes anthocyanin accumulation and enhances volatile aroma production in tomato fruits. Creative commons attribution 4.0 international license. *Horticulture Research, 6*, 22. https://doi.org/10.1038/s41438-018-0098-y (Accessed on 18 December 2019).

80. Jinturkar, K. A., Rathi, M. N., & Misra, A., (2011). Gene delivery using physical methods. In: Misra, A., (ed.), *Challenges in Delivery of Therapeutic Genomics and Proteomics* (pp. 83–126). Elsevier, Inc.: London, U.K. https://doi.org/10.1016/B978-0-12-384964-9.00003-7 (Accessed on 18 December 2019).
81. Jisha V., Dampanabolina, L., Vadassery, J., Mithöfer, A., Kappara, S., & Ramanan, R., (2015). Overexpression of an AP2/ERF type transcription factor OsEREBP1 confers biotic and abiotic stress tolerance in rice. Creative commons attribution 4.0 international license. *PLoS One, 10*(6), e0127831. doi: 10.1371/journal.pone.0127831.
82. Joersbo, M., & Brunstedt, J., (1996). Electroporation and transgenic plant production. In: Lynch, P. T., & Davey, M. R., (eds.), *Electrical Manipulation of Cells* (pp. 201–222). Springer: Boston, Massachusetts.
83. Kaeppler, H. F., Somers, D. A., Rines, H. W., & Cockburn, A. F., (1992). Silicon carbide fiber-mediated stable transformation of plant cells. *Theoretical and Applied Genetics, 84*, 560–566.
84. Karki, S., Rizal, G., & Quick, W. P., (2013). Improvement of photosynthesis in rice (*Oryza sativa* L.) by inserting the C4 pathway. Creative commons attribution 2.0 generic license. *Rice, 6*(1), 28. doi: 10.1186/1939-8433-6-28.
85. Kasuga, M., Miura, S., Shinozaki, K., & Yamaguchi, S., (2001). A combination of the *Arabidopsis* DREB1A gene and stress-inducible *rd29A* promoter improved drought- and low-temperature stress tolerance in tobacco by gene transfer. *Plant and Cell Physiology, 45*(3), 346–350. https://doi.org/10.1093/pcp/pch037 (Accessed on 18 December 2019).
86. Kebeish, R., Niessen, M., Thiruveedhi, K., Bari, R., Hirsch, H. J., Rosenkranz, R., Stäbler, N., Schönfeld, B., Kreuzaler, F., & Peterhänsel, C., (2007). Chloroplastic photorespiratory bypass increases photosynthesis and biomass production in *Arabidopsis thaliana*. *Nature Biotechnology, 25*, 593–599.
87. Khamis, G., Mona, A. A., Gihan, M. H., & Sadik, A. S., (2009). Application of laser microbeam cell surgery and *Agrobacterium-*

mediated gene transformation systems in melon (*Cucumis melo* L.). *Pakistan Journal of Biotechnology*, *6*(1), 45–54.

88. Khodakovskaya, M., McAvoy, R., Peters, J., Wu, H., & Li, Y., (2006). Enhanced cold tolerance in transgenic tobacco expressing a chloroplast omega-3 fatty acid desaturase gene under the control of a cold-inducible promoter. *Planta*, *223*(5), 1090–1100.

89. Kim, J. C., Lee, S. H., Cheong, Y. H., Yoo, C. M., Lee, S. I., Chun, H. J., Yun, D. J., Hong, J. C., Lee, S. Y., Lim, C. O., & Cho, M. J., (2001). A novel cold-inducible zinc finger protein from soybean, SCOF-1, enhances cold tolerance in transgenic plants. *The Plant Journal*, *25*(3), 247–251. https://doi.org/10.1046/j.1365-313x.2001.00947.x (Accessed on 18 December 2019).

90. Kim, M. H., Sasaki, K., & Imai, R., (2009). Cold shock domain protein 3 regulates freezing tolerance in *Arabidopsis thaliana*. *Journal of Biological Chemistry*, *284*(35), 23454–23460. doi: 10.1074/jbc.M109.025791.

91. Köhler, I. H., Ruiz-Vera, U. M., VanLoocke, A., Thomey, M. L., Clemente, T., Long, S. P., Ort, D. R., & Bernacchi, C. J., (2017). Expression of cyanobacterial FBP/SBPase in soybean prevents yield depression under future climate conditions. *Journal of Experimental Botany*, *68*(3), 715–726. https://doi.org/10.1093/jxb/erw435 (Accessed on 18 December 2019).

92. Kong, W., Liu, F., Zhang, C., Zhang, J., & Feng, H., (2016). Non-destructive determination of Malondialdehyde (MDA) distribution in oilseed rape leaves by laboratory scale NIR hyperspectral imaging. *Scientific Reports*, *6*, 35393.

93. Kromdijk, J., Glowacka, K., Leonelli, L., Gabilly, S. T., Iwai, M., Niyogi, K. K., & Long, S. P., (2016). Improving photosynthesis and crop productivity by accelerating recovery from photoprotection. *Science*, *354*(6314), 857–861. doi: 10.1126/science.aai8878.

94. Kumar, R., & Jain, N., (2017). Genetically modified mustard-dhara mustard hybrid-11. *Biotech Articles*. https://www.biotecharticles.com/Biotech-Research-Article/Genetically-Modified-Mustard-Dhara-Mustard-Hybrid-11-3986.html (Accessed on 18 December 2019).

95. Lai, H., He, J., Engle, M., Diamond, M. S., & Chen, Q., (2011). Robust production of virus-like particles and monoclonal antibodies with geminiviral replicon vectors in lettuce. *Plant Biotechnology Journal*, 10(1), 95–104. https://doi.org/10.1111/j.1467-7652.2011.00649.x

(Accessed on 18 December 2019).

96. Le Page, M., (2016). *Trials Planned for GM Super Wheat That Boosts Harvest by 20%. New Scientist*. https://www.newscientist.com/article/2111377-trials-planned-for-gm-superwheat-that-boosts-harvest-by-20/ (Accessed on 18 December 2019).
97. Ledford, H., (2014). Brazil considers transgenic trees. *Nature, 512*, 357. doi: 10.1038/512357a.
98. Lee, C., Kim, H. H., Choi, K. M., Chung, K. W., Choi, Y. K., Jang, M. J., Kim, T. S., et al., (2011). Murine immune responses to a *Plasmodium vivax*-derived chimeric recombinant protein expressed in *Brassica napus*. *Malaria Journal*, *10*, 106. doi: 10.1186/1475-2875-10-106.
99. Leite, M. L., Sampaio, K. B., Costa, F. F., Franco, O. L., Dias, S. C., & Cunha, N. B., (2019). Molecular farming of antimicrobial peptides: Available platforms and strategies for improving protein biosynthesis using modified virus vectors. Creative commons attribution 4.0 international license. *Anais da Academia Brasileira de Ciências, 91*(1). http://dx.doi.org/10.1590/0001-3765201820180124 (Accessed on 18 December 2019).
100. Lomonossoff, G. P., (1995). Pathogen-derived resistance to plant viruses. *Annual Review of Phytopathology*, *33*, 323–343.
101. Low, L. Y., Yang, S. K., Kok, D. X. A., Ong-Abdullah, J., Tan, N. P., & Lai, K. S., (2018). Transgenic plants: Gene constructs, vector and transformation method. In: Celik, O. (ed.), *New Visions in Plant Science*IntechOpen. doi: 10.5772/intechopen.79369.
102. Loza-Rubio, E., Rojas-Anaya, E., López, J., Olivera-Flores, M. T., Gómez-Lim, M., & Tapia-Pérez, G., (2012). Induction of a protective immune response to rabies virus in sheep after oral immunization with transgenic maize, expressing the rabies virus glycoprotein. *Vaccine*, *30*(37), 5551–5556. https://doi.org/10.1016/j.vaccine.2012.06.039 (Accessed on 18 December 2019).
103. Lucca, P., Hurrell, R., & Potrykus, I., (2002). Fighting iron deficiency anemia with iron-rich rice. *Journal of the American College of Nutrition*, *21*(3), 184S–190S. doi: 10.1080/07315724.2002.10719264.
104. Ma, Q., Dai, X., Xu, Y., Guo, J., Liu, Y., Chen, N., Xiao, J., Zhang, D., Zhang, X. Z., & Chong, K., (2009). Enhanced tolerance to chilling stress in *OsMYB3R-2* transgenic rice is mediated by alteration in cell cycle and ectopic expression of stress genes. *Plant Physiology*, *150*(1),

244–256. https://doi.org/10.1104/pp.108.133454 (Accessed on 18 December 2019).

105. Madsen, K. H., & Streibig, J. C., (2019). *Benefits and Risks of the Use of Herbicide-Resistant Crops*. http://www.fao.org/3/y5031e/y5031e0i.htm (Accessed on 18 December 2019).

106. Majumder, S., Datta, K., Sarkar, C., Saha, S. C., & Datta, S. K., (2018). The development of *Macrophina phaseolina* (fungus) resistant and glufosinate (herbicide) tolerant transgenic jute. Creative commons attribution 4.0 international license. *Frontiers in Plant Science*, *9*, 920. https://doi.org/10.3389/fpls.2018.00920 (Accessed on 18 December 2019).

107. Margolin, E., Chapman, R., Meyers, A. E., Van Diepen, M. T., Ximba, P., Hermanus, T., Crowther, C., Weber, B., Morris, L., Williamson, A. L., & Rybicki, E. P., (2019). Production and immunogenicity of soluble plant-produced HIV-1 subtype C envelope gp140 immunogens. *Frontiers in Plant Science*, *10*, 1378. doi: 10.3389/fpls.2019.01378.

108. Margolin, E., Chapman, R., Williamson, A. L., Rybicki, E. P., & Meyers, A. E., (2018). Production of complex viral glycoproteins in plants as vaccine immunogens. Creative commons attribution 4.0 international license. *Plant Biotechnology Journal*, *16*(9), 1531–1545. https://doi.org/10.1111/pbi.12963 (Accessed on 18 December 2019).

109. Marillonnet, S., Thoeringer, C., Kandzia, R., Klimyuk, V., & Gleba, Y., (2005). Systemic *Agrobacterium tumefaciens*-mediated transfection of viral replicons for efficient transient expression in plants. *Nature Biotechnology*, *23*, 718–723.

110. Masani, M. Y. A., Noll, G. A., Parveez, G. K. A., Sambanthamurthi, R., & Prüfer, D., (2014). Efficient transformation of oil palm protoplasts by PEG-mediated transfection and DNA microinjection. Creative commons attribution 4.0 international license. *PLoS One*, 9(5), e96831. https://doi.org/10.1371/journal.pone.0096831 (Accessed on 18 December 2019).

111. McCabe, D., & Christou, P., (1993). Direct DNA transfer using electric discharge particle acceleration (ACCELL™ technology). *Plant Cell, Tissue and Organ Culture*, *33*, 227–236.

112. McKersie, B. D., Bowley, S. R., Harjanto, E., & Leprince, O., (1996). Water-deficit tolerance and field performance of transgenic alfalfa overexpressing superoxide dismutase. *Plant Physiology*, *111*, 1177–1181.

113. Mehta, R. A., Cassol, T., Li, N., Ali, N., Handa, A. K., & Mattoo, A. K., (2002). Engineered polyamine accumulation in tomato enhances phytonutrient content, juice quality, and vine life. *Nature Biotechnology, 20*, 613–618.

114. Mian, A., Oomen, R. J. F. J., Isayenkov, S., Sentenac, H., Maathuis, F. J. M., & Véry, A. A., (2011). Over-expression of an Na+- and K+-permeable HKT transporter in barley improves salt tolerance. *The Plant Journal, 68*(3), 468–479. https://doi.org/10.1111/j.1365-313X.2011.04701.x (Accessed on 18 December 2019).

115. Miles, C. M., & Wayne, M., (2008). Quantitative trait locus (QTL) analysis. *Nature Education, 1*(1), 208.

116. Mitra, J., (2001). Genetics and genetic improvement of drought resistance in crop plants. *Current Science, 80*(6), 758–763.

117. Mouagip. *Public Domain.* https://commons.wikimedia.org/wiki/File:Ti_plasmid.svg (Accessed on 18 December 2019).

118. Mukhopadhyay, A., Vij, S., & Tyagi, A. K., (2004). Overexpression of a zinc-finger protein gene from rice confers tolerance to cold, dehydration, and salt stress in transgenic tobacco. *Proceedings of the National Academy of Sciences of the United States of America, 101*(16), 6309–6314. https://doi.org/10.1073/pnas.0401572101 (Accessed on 18 December 2019).

119. Murata, N., Ishizaki-Nishizawa, O., Higashi, S., Hayashi, H., Tasaka, Y., & Nishida, I., (1992). Genetically engineered alteration in the chilling sensitivity of plants. *Nature, 356*, 710–713.

120. Nanjo, T., Kobayashi, M., Yoshiba, Y., Kakubati, Y., Yamaguchi-Shinozaki, K., 7 Shinozaki, K., (1999). Antisense suppression of proline degradation improves tolerance to freezing and salinity in *Arabidopsis thaliana. FEBS Letters, 461*(3), 205–210. https://doi.org/10.1016/S0014-5793(99)01451-9 (Accessed on 18 December 2019).

121. Narusaka, Y., Narusaka, M., Yamasaki, S., & Iwabuchi, M., (2012). Methods to transfer foreign genes to plants. In: Çiftçi, Y. O., (ed.), Creative commons attribution 3.0 license. *Transgenic Plants-Advances and Limitations*, IntechOpen. doi: 10.5772/32773.

122. National Academy of Sciences, National Research Council (US) Committee on Identifying and Assessing Unintended Effects of Genetically Engineered Foods on Human Health, (2004). *Safety of Genetically Engineered Foods: Approaches to Assessing Unintended*

Health Effects. National Academies Press: Washington D.C.

123. National Museum of American History, (2019). *ACCELL Gene Gun Prototype*. https://americanhistory.si.edu/collections/search/object/nmah_1165091 (Accessed on 18 December 2019).
124. Nature, (2019). *Biolistics*. https://www.nature.com/subjects/biolistics (Accessed on 18 December 2019).
125. Newell-McGloughlin, M., (2008). Nutritionally improved agricultural crops. *Plant Physiology*, *147*(3), 939–953. doi: 10.1104/pp.108.121947.
126. Norero, D., (2018). GMO crops have been increasing yield for 20 years, with more progress ahead. *Cornell Alliance for Science*. https://allianceforscience.cornell.edu/blog/2018/02/gmo-crops-increasing-yield-20-years-progress-ahead/ (Accessed on 18 December 2019).
127. Nykiforuk, C. L., Shewmaker, C., Harry, I., Yurchenko, O. P., Zhang, M., Reed, C., Oinam, G. S., Zaplachinski, S., Fidantsef, A., Boothe, J. G., & Moloney, M. M., (2012). High level accumulation of gamma linolenic acid (C18:3Δ6.9, 12 *cis*) in transgenic safflower (*Carthamus tinctorius*) seeds. *Transgenic Research*, *21*(2), 367–381.
128. Oxford Reference, (2019). *ColE1 Replicon*. https://www.oxfordreference.com/view/10.1093/oi/authority.20110803095623550 (Accessed on 18 December 2019).
129. Petaholmes, B., & Gallet, S. (2009) *Creative Commons Attribution = Share Alike 3.0 Unported License*. https://commons.wikimedia.org/wiki/File:Carotenoidsynthesis.svg (Accessed on 18 December 2019).
130. Pilon-Smits, E. A. H., Ebskamp, M. J. M., Paul, M. J., Jeuken, M. J. W., Weisbeek, P. J., & Smeekens, S. C. M., (1995). Improved performance of transgenic fructan-accumulating tobacco under drought stress. *Plant Physiology*, *107*, 125–130.
131. Pogrebnyak, N., Golovkin, M., Andrianov, V., Spitsin, S., Smirnov, Y., Egolf, R., & Koprowski, H., (2005). Severe acute respiratory syndrome (SARS) S protein production in plants: Development of recombinant vaccine. *Proceedings of the National Academy of Sciences of the United States of America*, *102*(25), 9062–9067. https://doi.org/10.1073/pnas.0503760102 (Accessed on 18 December 2019).
132. Pollegioni, L., Schonbrunn, E., & Siehl, D., (2011). Molecular basis of glyphosate resistance: Different approaches through protein engineering. *FEBS Journal*, *278*(16), 2753–2766. https://doi.org/10.1111/j.1742-4658.2011.08214.x (Accessed on 18 December

2019).

133. Poolcharoen, W., Bhoo, S. H., Lai, H., Ma, J., Arntzen, C. J., Chen, Q., & Mason, H. S., (2011). Expression of an immunogenic Ebola immune complex in *Nicotiana benthamiana. Plant Biotechnology Journal, 9*(7), 807–816. https://doi.org/10.1111/j.1467-7652.2011.00593.x (Accessed on 18 December 2019).
134. Pseudoanas, *Public Domain.* https://ugeneko.exblog.jp/2681997/ (Accessed on 18 December 2019).
135. Qian, W., Xiao, B., Wang, L., Hao, X., Yue, C., Cao, H., Wang, Y., Li, N., Yu, Y., Zeng, J., Yang, Y., & Wang, X., (2018). *CsINV5*, a tea vacuolar invertase gene enhances cold tolerance in transgenic *Arabidopsis. BMC Plant Biology, 18*(228). https://doi.org/10.1186/s12870-018-1456-5 (Accessed on 18 December 2019).
136. Quimio, C. A., Torrizo, L. B., Setter, T. L., Ellis, M., Grover, A., Abrigo, E. M., Oliva, N. P., Ella, E. S., Carpena, A. L., Ito, O., Peacock, W. J., Dennis, E., & Dattal, S. K., (2000). Enhancement of submergence tolerance in transgenic rice overproducing pyruvate decarboxylase. *Journal of Plant Physiology, 156*(4), 516–521. https://doi.org/10.1016/S0176-1617(00)80167-4 (Accessed on 18 December 2019).
137. RachelBrooks15. *Creative Commons Attribution-Share Alike 4.0 International License.* https://commons.wikimedia.org/wiki/File:Gene_Gun.svg (Accessed on 18 December 2019).
138. Rakoczy-Trojanowska, M., (2002). Alternative methods of plant transformation-a short review. *Cellular and Molecular Biology Letters, 7*, 849–858.
139. Rao, A. Q., Ali, M. A., Khan, M. A. U., Bajwa, K. S., Iqbal, A., Iqbal, T., Shahid, A. A., Nasir, I. A., & Husnain, T., (2016). In: Abdurakhmonov, I. Y., (ed.), *Science Behind Cotton Transformation. Cotton Research.* IntechOpen. doi: 10.5772/64888.
140. Rashid, A. H. A., & Lateef, D. D., (2016). Novel techniques for gene delivery into plants and its applications for disease resistance in crops. *American Journal of Plant Sciences, 7*, 181–193. http://dx.doi.org/10.4236/ajps.2016.71019 (Accessed on 18 December 2019).
141. Rivera, A. L., Gómez-Lim M., Fernández, F., & Loske, A. M., (2014). Genetic transformation of cells using physical methods. *Journal of Genetic Syndromes and Gene Therapy, 5*(4). doi: 10.4172/2157-7412.1000237.

142. Rizal, G., Karki, S., Thakur, V., Chatterjee, J., Coe, R. A., Wanchana, S., & Quick, W. P., (2012). Towards a C_4 rice. *Asian Journal of Cell Biology*, 1–19. doi: 10.3923/ajcb.2012.

143. Ronald, P. *The Ronald Laboratory, Crop Genetics Innovation and Scientific Literacy: Engineering Submergence Tolerance in Rice*. https://cropgeneticsinnovation.ucdavis.edu/engineering-submergence-tolerance-rice (Accessed on 18 December 2019).

144. Roxas, V. P., Smith, R. K., Allen, E. R., & Allen, R. D., (1997). Overexpression of glutathione S-transferase/glutathioneperoxidase enhances the growth of transgenic tobacco seedlings during stress. *Nature Biotechnology*, *15*, 988–991.

145. Roy, S. J., Negrão, S., & Tester, M., (2014). Salt resistant crop plants. *Current Opinion in Biotechnology*, *26*, 115–124. https://doi.org/10.1016/j.copbio.2013.12.004 (Accessed on 18 December 2019).

146. Ruhlman, T., Ahangari, R., Devine, A., Samsam, M., & Daniell, H., (2007). Expression of cholera toxin B-proinsulin fusion protein in lettuce and tobacco chloroplasts-oral administration protects against development of insulitis in non-obese diabetic mice. *Plant Biotechnology Journal*, *5*(4), 495–510. https://doi.org/10.1111/j.1467-7652.2007.00259.x (Accessed on 18 December 2019).

147. Saejung, W., Fujiyama, K., Takasaki, T., Ito, M., Hori, K., Malasit, P., Watanabe, Y., Kurane, I., & Seki, T., (2007). Production of dengue 2 envelope domain III in plant using TMV-based vector system. *Vaccine*, *25*(36), 6646–6654. doi: 10.1016/j.vaccine.2007.06.029.

148. Sakamoto, A., & Murata, A., (1998). Metabolic engineering of rice leading to biosynthesis of glycinebetaine and tolerance to salt and cold. *Plant Molecular Biology*, *38*(6), 1011–1019.

149. Sanford, J. C., & Johnston, S. A., (1985). The concept of parasite-derived resistance-deriving resistance genes from the parasite's own genome. *Journal of Theoretical Biology*, *113*, 395–405.

150. Sanghera, G. S., Wani, S. H., Hussain, W., & Singh, N. B., (2011). Engineering cold stress tolerance in crop plants. *Current Genomics*, *12*(1), 30–43. doi: 10.2174/138920211794520178.

151. Sarker, S. R., & Takeoka, S., (2018). Amino acid-based liposomal assemblies: Intracellular plasmid DNA delivery nanoparticles. Creative commons attribution 4.0 international license. *Journal of Nanomedicine*, *2*, 1008.

152. Serres, R., Stang, E., McCabe, D., Russell, D., Mahr, D., & McCown, B., (1992). Gene transfer using electric discharge particle bombardment and recovery of transformed cranberry plants. *Journal of the American Society for Horticultural Science*, *117*(1), 174–180. https://doi.org/10.21273/JASHS.117.1.174 (Accessed on 18 December 2019).

153. Sévenier, R., Hall, R. D., Van Der Meer, I. M., Hakkert, H. J. C., Van Tunen, A. J., & Koops, A. J., (1998). High level fructan accumulation in a transgenic sugar beet. *Nature Biotechnology*, *16*, 843–846.

154. Shelton, A. M., Hossain, M. J., Paranjape, V., Azad, A. K., Rahman, M. L., Khan, A. S. M. M. R., Prodhan, M. Z. H., Rashid, M. A., Majumder, R., Hossain, M. A., Hussain, S. S., Huesing, J. E., & McCandless, L., (2018). Bt eggplant project in Bangladesh: History, present status, and future direction. *Frontiers in Bioengineering and Biotechnology*, *6*, 106. doi: 10.3389/fbioe.2018.00106.

155. Shepherd, D. N., Dugdale, B., Martin, D. P., Varsani, A., Lakay, F. M., Bezuidenhout, M. E., Monjane, A. L., Thomson, J. A., Dale, J., & Rybicki, E. P., (2014). Inducible resistance to maize streak virus. *PLoS One*, *9*(8), e105932. https://doi.org/10.1371/journal.pone.0105932 (Accessed on 18 December 2019).

156. Shepherd, D. N., Mangwende, T., Martin, D. P., Bezuidenhout, M., Kloppers, F. J., Carolissen, C. H., Monjane, A. L., Rybicki, E. P., & Thomson, J. A., (2007). Maize streak virus-resistant transgenic maize: A first for Africa. *Plant Biotechnology Journal*, *5*(6), 759–767. https://doi.org/10.1111/j.1467-7652.2007.00279.x (Accessed on 18 December 2019).

157. Shinozaki, K., & Yamaguchi-Shinozaki, K., (2007). Gene networks involved in drought stress response and tolerance. *Journal of Experimental Botany*, *58*(2), 221–227. https://doi.org/10.1093/jxb/erl164 (Accessed on 18 December 2019).

158. Shrivastava, P., & Kumar, R., (2015). Soil salinity: A serious environmental issue and plant growth promoting bacteria as one of the tools for its alleviation. *Saudi Journal of Biological Sciences*, *22*(2), 123–131. doi: 10.1016/j.sjbs.2014.12.001.

159. Slewinski, T. L., Anderson, A. A., Zhang, C., & Turgeon, R., (2012). *Scarecrow* plays a role in establishing Kranz anatomy in maize leaves. *Plant and Cell Physiology*, *53*(12), 2030–2037. https://doi.org/10.1093/pcp/pcs147 (Accessed on 18 December 2019).

160. South, P. F., Cavanagh, A. P., Liu, H. W., & Ort, D. R., (2019). Synthetic

glycolate metabolism pathways stimulate crop growth and productivity in the field. Creative commons attribution 4.0 international license. *Science*, *363*(6422), eaat9077. doi: 10.1126/science.aat9077.

161. Su, C. F., Wang, Y. C., Hsieh, T. H., Lu, C. A., Tseng, T. H., & Yu, S. M., (2010). A novel MYBS3-dependent pathway confers cold tolerance in rice. *Plant Physiology*, *153*, 145–158. https://doi.org/10.1104/pp.110.153015 (Accessed on 18 December 2019).

162. Suzuki, K., Kaminuma, O., Yang, L., Takai, T., Mori, A., Umezu-Goto, M., Ohtomo, T., Ohmachi, Y., Noda, Y., Hirose, S., Okumura, K., Ogawa, H., Takada, K., Hirasawa, M., Hiroi, T., & Takaiwa, F., (2011). Prevention of allergic asthma by vaccination with transgenic rice seed expressing mite allergen: Induction of allergen-specific oral tolerance without bystander suppression. *Plant Biotechnology Journal*, *9*(9), 982–990. https://doi.org/10.1111/j.1467-7652.2011.00613.x (Accessed on 18 December 2019).

163. Tamminen, I., Mäkela, P., Heino, P., & Palva, E. T., (2008). Ectopic expression of *ABI3* gene enhances freezing tolerance in response to abscisic acid and low temperature in *Arabidopsis thaliana*. *The Plant Journal*, *25*(1), 1–8. https://doi.org/10.1111/j.1365-313X.2001.00927.x (Accessed on 18 December 2019).

164. Toenniessen, G. H., (2002). Crop genetic improvement for enhanced human nutrition. *The Journal of Nutrition*, *132*(9), 2943S–2946S. https://doi.org/10.1093/jn/132.9.2943S (Accessed on 18 December 2019).

165. Tripathi, L., Atkinson, H., Roderick, H., Kubiriba, J., & Tripathi, J. N., (2017). Genetically engineered bananas resistant to *Xanthomonas* wilt disease and nematodes. *Food and Energy Security*, *6*(2), 37–47. https://doi.org/10.1002/fes3.101 (Accessed on 18 December 2019).

166. Tzotzos, G. T., Head, G. P., & Hull, R., (2009). *Genetically Modified Plants: Assessing Safety and Managing Risk*. Academic Press: London, U.K.

167. Umezawa, T., Okamoto, M., Kushiro, T., Nambara, E., Oono, Y., Seki, M., Kobayashi, M., Koshiba, T., Kamiya, Y., & Shinozaki, K., (2006). CYP707A3, a major ABA 8'-hydroxylase involved in dehydration and rehydration response in *Arabidopsis thaliana*. *The Plant Journal*, *46*(2), 171–182. https://doi.org/10.1111/j.1365-313X.2006.02683.x (Accessed on 18 December 2019).

168. University of Oxford, (2017). Breakthrough in efforts to 'supercharge'

rice and reduce world hunger. *News and Events*. http://www.ox.ac.uk/news/2017-10-19-breakthrough-efforts-supercharge-rice-and-reduce-world-hunger# (Accessed on 18 December 2019).

169. Ursin, V. M., (2003). Modification of plant lipids for human health: development of functional land-based omega-3 fatty acids. *The Journal of Nutrition, 133*(12), 4271–4274. https://doi.org/10.1093/jn/133.12.4271 (Accessed on 18 December 2019).
170. USDA. *Public Domain.* https://commons.wikimedia.org/wiki/File:Canola.jpg (Accessed on 18 December 2019).
171. Uvarova, E. A., Belavin, P. A., Permyakova, N. V., Zagorskaya, A. A., Nosareva, O. V., Kakimzhanova, A. A., & Deineko, E. V., (2013). Oral immunogenicity of plant-made *Mycobacterium tuberculosis* ESAT6 and CFP10. *BioMed Research International*, 316304. doi: 10.1155/2013/316304.
172. Vannini, C., Locatelli, F., Bracale, M., Magnani, E., Marsoni, M., Osnato, M., Mattana, M., Baldoni, E., & Coraggio, I., (2003). Overexpression of the rice *Osmyb4* gene increases chilling and freezing tolerance of *Arabidopsis thaliana* plants. *The Plant Journal, 37*(1), 115–127. https://doi.org/10.1046/j.1365-313X.2003.01938.x (Accessed on 18 December 2019).
173. Vogel, J. T., Zarka, D. G., Van Buskirk, H. A., Fowler, S. G., & Thomashow, M. F., (2004). Roles of the CBF2 and ZAT12 transcription factors in configuring the low temperature transcriptome of arabidopsis. *The Plant Journal*, *41*(2), 195–211. https://doi.org/10.1111/j.1365-313X.2004.02288.x (Accessed on 18 December 2019).
174. Wang, C., Guo, L., Li, Y., & Wang, Z., (2012). Systematic comparison of C3 and C4 plants based on metabolic network analysis. Creative commons attribution 2.0 generic license. *BMC Systems Biology, 6*(2), S9. doi: 10.1186/1752-0509-6-S2-S9.
175. Wang, Y., Cheng, X., Shan, Q., Zhang, Y., Liu, J., Gao, C., & Qiu, J. L., (2014). Simultaneous editing of three homeoalleles in hexaploidy bread wheat confers heritable resistance to powdery mildew. *Nature Biotechnology*, *9*, 947–951. doi: 10.1038/nbt.2969.
176. Wang, Y., Ying, J., Kuzma, M., Chalifoux, M., Sample, A., McArthur, C., Uchacz, T., Sarvas, C., Wan, J., Dennis, D. T., McCourt, P., & Huang, Y., (2005). Molecular tailoring of farnesylation for plant drought tolerance and yield protection. *The Plant Journal*, *43*(3), 413–

424. https://doi.org/10.1111/j.1365-313X.2005.02463.x (Accessed on 18 December 2019).

177. Wheeler, R. A: (2006) *Creative Commons Attribution-Share Alike 3.0 Unported License.* https://commons.wikimedia.org/wiki/File:Electroporation_Cuvettes.jpg; B: (2007) *Creative Commons Attribution-Share Alike 3.0 Unported License.* https://commons.wikimedia.org/wiki/File:Electroporation_Diagram.png (Accessed on 18 December 2019).

178. Wilkerson, J., (2015). *Why Roundup Ready Crops Have Lost Their Allure.* Science in the news. http://sitn.hms.harvard.edu/flash/2015/roundup-ready-crops/ (Accessed on 18 December 2019).

179. Xmort. *Public Domain.* https://commons.wikimedia.org/wiki/File:Genegun.jpg (Accessed on 18 December 2019).

180. Xu, J., Xing, X. J., Tian, Y. S., Peng, R. H., Xue, Y., Zhao, W., & Yao, Q. H., (2015). Transgenic arabidopsis plants expressing tomato glutathione s-transferase showed enhanced resistance to salt and drought stress. Creative commons attribution 4.0 international license. *PLoS One*, *10*(9), e0136960. doi: 10.1371/journal.pone.0136960.

181. Xu, K., Xu, X., Fukao, T., Canlas, P., Maghirang-Rodriguez, R., Heuer, S., Ismail, A. M., Bailey-Serres, J., Ronald, P. C., & Mackill, D. J., (2006). *Sub1A* is an ethylene-response-factor-like gene that confers submergence tolerance in rice. *Nature*, *442*, 705–708.

182. Yang, M., Sun, H., Lai, H., Hurtado, J., & Chen, Q., (2017). Plant-produced Zika virus envelope protein elicits neutralizing immune responses that correlate with protective immunity against Zika virus in mice. *Plant Biotechnology Journal*, *16*(2), 572–580. https://doi.org/10.1111/pbi.12796 (Accessed on 18 December 2019).

183. Yikrazuul, *Public Domain.* https://commons.wikimedia.org/wiki/File:Violaxanthin_cycle.png (Accessed on 18 December 2019).

184. Young, T. E., Geisler-Lee, J., & Gallie, D. R., (2004). Senescence-induced expression of cytokinin reverses pistil abortion during maize flower development. *The Plant Journal*, *38*(6), 910–922. https://doi.org/10.1111/j.1365-313X.2004.02093.x (Accessed on 18 December 2019).

185. Zaidi, A. S., & Mansoor, S., (2017). Viral vectors for plant genome engineering. *Frontiers in Plant Science*, *8*, 539. doi: 10.3389/fpls.2017.00539.

186. Zechner, E. L., Lang, S., & Schildbach, J. F., (2012). Assembly and mechanisms of bacterial Type IV secretion machines. Creative commons attribution 3.0 unported license. *Philosophical Transactions of the Royal Society B Biological Sciences, 367*(1592), 1073–1087. doi: 10.1098/rstb.2011.0207.

187. Zhang, H. X., Hodson, J. N., Williams, J. P., & Blumwald, E., (2001). Engineering salt-tolerant *Brassica* plants: Characterization of yield and seed oil quality in transgenic plants with increased vacuolar sodium accumulation. *Proceedings of the National Academy of Sciences of the United States of America, 98*(22), 12832–12836. https://doi.org/10.1073/pnas.231476498 (Accessed on 18 December 2019).

188. Zhang, Y. J., Yang, J. S., Guo, S. J., Meng, J. J., Zhang, Y. L., Wan, S. B., He, Q. W., & Li, X. G., (2011). Over-expression of the *Arabidopsis CBF1* gene improves resistance of tomato leaves to low temperature under low irradiance. *Plant Biology, 13*(2), 362–367. https://doi.org/10.1111/j.1438-8677.2010.00365.x (Accessed on 18 December 2019).

189. Zhu, B., Xiong, A. S., Peng, R. H., Xu, J., Jin, X. F., Meng, X. R., & Yao, Q. H., (2010). Over-expression of *ThpI* from *Choristoneura funiferana* enhances tolerance to cold in Arabidopsis. *Molecular Biology Reports, 37*, 961. https://doi.org/10.1007/s11033-009-9759-0 (Accessed on 18 December 2019).

Chapter 4

Genome Editing in Plants

CONTENTS

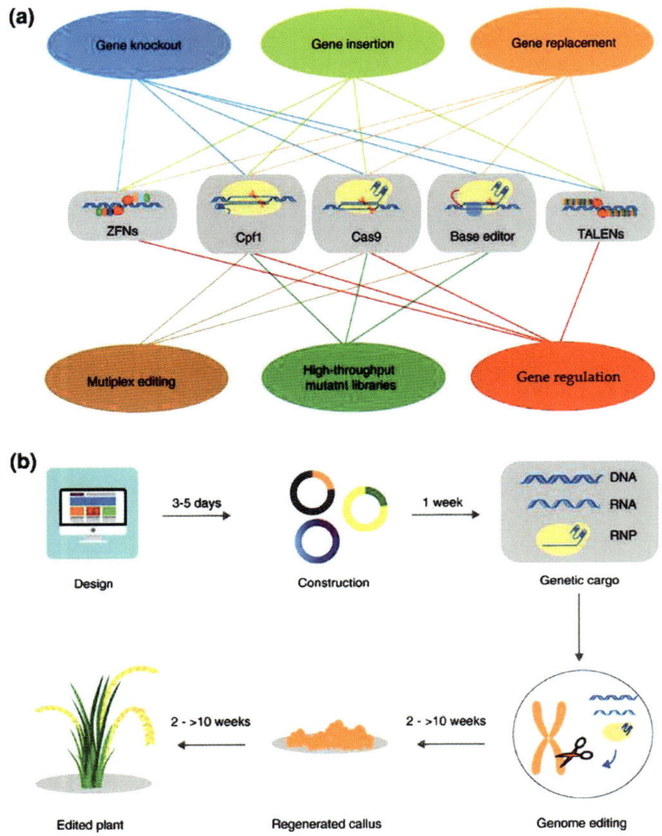

(From Zhang, Y., Massel, K., Godwin, I. D., & Gao, C. (2018). Genome Biology 19, p. 210, Creative Commons License).

Genetically modified crops, as discussed in the previous chapter, express beneficial phenotypes due to the introduction of an exogenous gene in their genome. While these crops hold great potential in improving food production and in providing cheaper alternative for generating therapeutic compounds, there are concerns regarding the effect of transgenic crops on human health and the environment. These concerns prompted the framing of government regulations on biosafety. On the other hand, such restrictions made it more costly to cultivate transgenic crops so that only a small number of these are adopted for large-scale cultivation (Zhang et al., 2018).

Another method of generating new varieties is by mutation breeding. Traditional breeders utilize chemical compounds and irradiation to induce random mutations, which can potentially introduce beneficial and heritable mutations into plant genomes. However, the non-specificity of the mutation process and the simultaneous mutation of large genomic fragments make the identification of the mutation of interest a long and labor-intensive process. Additionally, random mutagenesis is not an efficient process for improving the trait of polyploid crops due to their multiple genomes (Mao et al., 2019).

The drawbacks of genetic engineering and mutation breeding led to the development of gene or genome editing techniques that enable the precise and efficient modification of certain locus or loci in the genome. These include the use of zinc-finger nucleases (ZFNs), transcription activator-like effector nucleases (TALENs), and type II clustered regularly interspaced short palindromic repeats (CRISPR)/CRISPR-associated protein (CAS) systems (see opening image).

4.1. INDUCTION AND REPAIR OF DOUBLE-STRANDED BREAKS (DSB)

The previously mentioned genome editing techniques facilitate the generation of double-stranded breaks (DSBs) along the DNA in a site-specific manner. The DSB can be within the protein coding sequence of a gene, non-protein coding genes (e.g., microRNA or long non-coding RNA) or regulatory elements (e.g., promoter, silencer, enhancer) (see Figure 4.1a, b, c). In other cases, large segments of the chromosomes may be deleted (see Figure 4.1d). In eukaryotic cells, these DSBs can be repaired either by the error-prone, non-homologous end-joining (NHEJ) pathway or the error-free homology-directed repair (HDR) pathway (Mao et al., 2019; Malzahn et al., 2017; Zhang et al., 2018).

NHEJ involves the non-homologous re-ligation of broken ends (see Figure 4.2). This is the most common DNA repair mechanism in plants. This repair process is error prone, which may result in small insertions, deletions, and/or rearrangements. As a result, mutations can be introduced which may affect gene function (Petolino, 2015). Plant genomes that were modified via NHEJ that is induced by genome editing tools may carry the targeted mutation but lack any transgenes. This occurs if the ZFN, TALEN, or CRISPR/Cas9 expression constructs are not integrated into the genome and are degraded after transient expression.

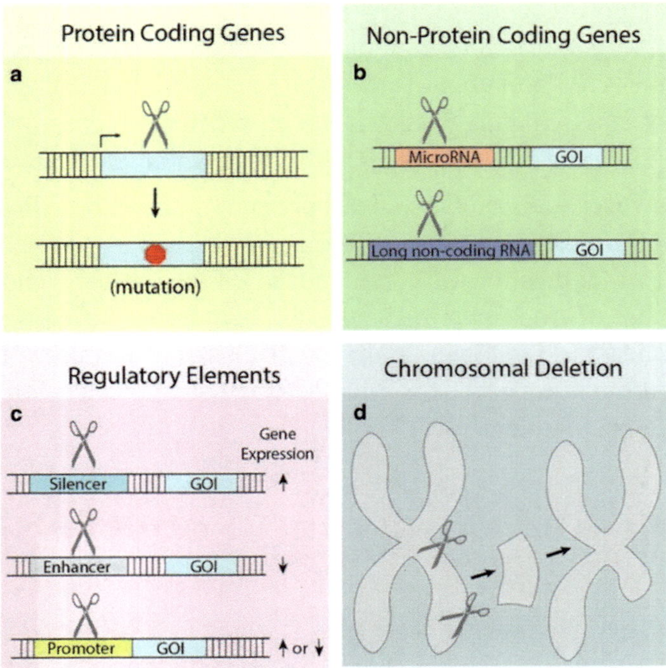

Figure 4.1: *Sequence-specific nucleases can induce double-stranded breaks (DSBs) within (a) protein coding genes, (b) non-protein coding genes or (c) regulatory elements. (d) The introduced DSBs may also result in the deletion of large chromosomal segments.*

Source: Malzahn, Lowder, and Qi (2017).

On the other hand, HDR utilizes donor DNA sequences, which are homologous to those flanking the DSB, as templates for synthesis-dependent strand annealing (see Figure 4.2). The homologous sequences may be designed to contain desired sequences that can be copied into the targeted cleavage site. The predetermined sequence may contain single to few base-pair modifications resulting in gene editing or may have a complete transgene expression cassette that can be integrated in a site-specific manner (Petolino, 2015).

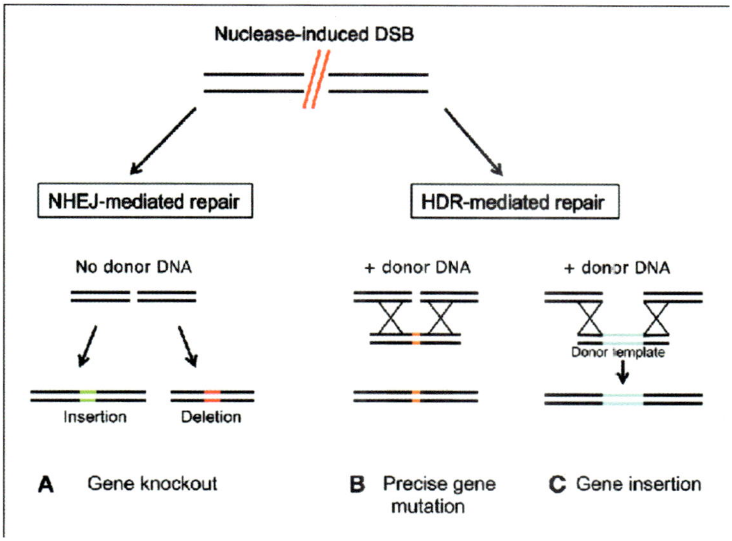

Figure 4.2: *A double-stranded break (DSB) of the target DNA that is induced by nucleases can be repaired by either non-homologous end-joining (NHEJ) or homology-directed repair (HDR). (A) The broken ends are joined together inaccurately in NHEJ-mediated repair, which is an error prone process resulting to either insertion or deletion of random nucleotides. This may lead to gene knockout or loss of function of the gene. HDR involves a donor DNA template with regions that are homologous to the sequences flanking the DSB. The DSB is repaired via homologous recombination which can introduce (B) precise gene modification or (C) gene insertion.*

Source: Limera, Sabbadini, Sweet, and Mezzetti (2017).

4.2. ZINC-FINGER NUCLEASES (ZFNS)

ZFNs are artificial endonucleases made by fusing DNA recognition domains that are zinc-finger based with the DNA-cleavage domain of the *Fok*1 restriction enzyme. Each zinc finger (ZF) recognizes and binds three nucleotides, and the fingers are grouped together to bind specific DNA sequences (Zhang et al., 2018) (see Figure 4.3). The *Fok*1 cleavage domain must form a dimer (specifically, a homodimer) before it can cut DNA because a monomer is not active. This means that if a zinc-finger domain that is joined to a monomeric cleavage domain binds to a specific DNA sequence, no cleavage occurs. However, if another ZFN binds a neighboring sequence in the opposite strand, the two ZFNS dimerize and facilitate cleavage (Carroll, 2011). This interaction between two ZFNs requires appropriate spacing and longer recognition sequences resulting in higher level of specificity. *Fok*1 variants can also be designed to require heterodimerization, which further enhances sequence specificity and minimizes off-site cleavage (Petolino, 2015).

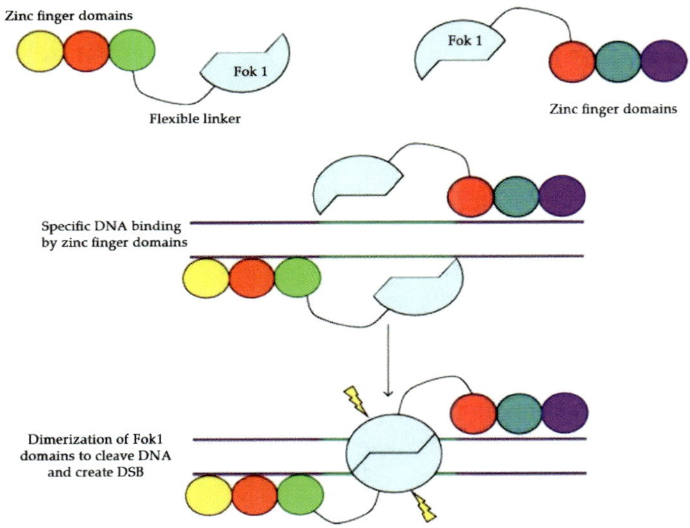

Figure 4.3: *A zinc-finger nuclease (ZFN) is designed by joining DNA-binding zinc-fingers to the DNA-cleavage domain of Fokl endonuclease. Each zinc finger, represented by a circle, recognizes three nucleotides and so assembling three or four zinc finger motifs produces a sequence-specific DNA-binding domain. ZFNs function in pairs because Fokl nuclease activity requires dimerization. The zinc finger-binding domain brings two Fokl units together in the right orientation over the target sequence. This induces Fokl dimerization and cleavage of the target sequence ensues.*

Source: Papaioannou, Simons, and Owen (2012).

ZFNs can be designed to bind and cleave any target DNA sequence to create DSBs and initiate the process to repair them. NHEJ-mediated repair may result to targeted mutagenesis while HDR may lead to gene editing or transgene integration (see Figure 4.4).

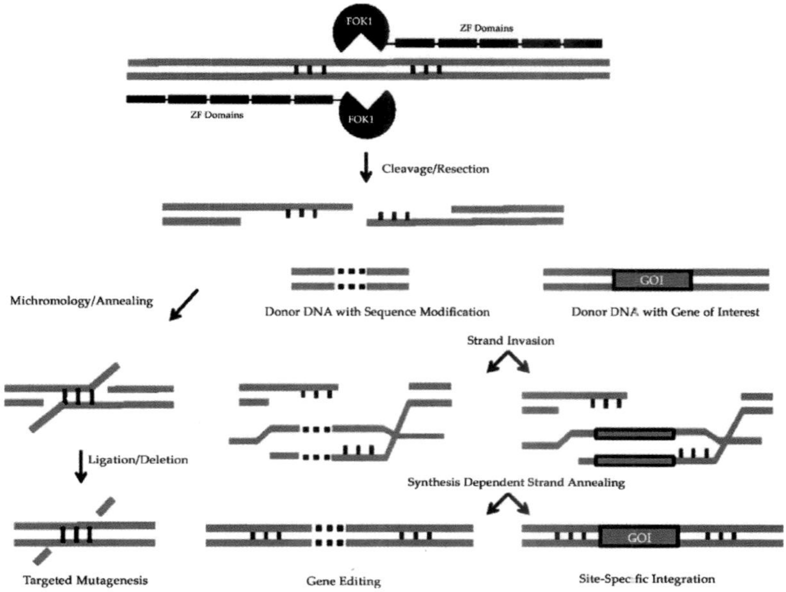

Figure 4.4. *A pair of zinc-finger nucleases (ZFNs) recognize and bind specific DNA sequences via their zinc finger (ZF) domains. The dimerization of the Fok1 nuclease domains of the ZFN pair makes them cleave the DNA to create a double-strand break (DSB), which in turn initiates the DNA repair processes. The inaccurate non-homologous end-joining (NHEJ) can result in targeted mutagenesis. The homology-directed repair (HDR) occurs in the presence of a donor DNA with sequences homologous to the regions flanking the DSB. The donor DNA may be designed to carry a specific sequence modification, which can result to gene editing, while a donor DNA carrying a gene of interest may lead to site-specific integration.*

Source: Petolino (2015).

Expression of ZFN is accomplished by constructing a plasmid vector containing the ZFN coding sequences under the control of a certain promoter. This is then delivered into the target plant cell by *A. tumefaciens*-mediated transformation. In other cases, other transformation methods such as polyethylene glycol (PEG) delivery into protoplasts, microparticle bombardment, WHISKERS™-mediated transformation, direct DNA transformation, and viral delivery have also been used successfully in plants. Apart from delivering active ZFNs, these transformation systems may also be used to deliver donor template DNA. It is critical that these reagents should be delivered at the appropriate stage of the cell cycle to ensure targeted cleavage and subsequent DNA repair (Carroll, 2011; Petolino, 2015).

4.2.1. Gene Targeting in Plants with Zinc-Finger Nucleases (ZFNs)

The first ZFNs were used as chimeric restriction endonucleases *in vitro*. Then ZFNs of known specificity were used on *Xenopus* oocytes with a synthetic, extrachromosomal substrate. These showed very high efficiency of cleavage and recombination. The first use of ZFN for genomic editing was on the *yellow* locus in the soma and germline of *Drosophila*. A pair of ZFNs have been designed to induce both targeted mutagenesis and targeted gene replacement (Carroll, 2011). ZFNs have now been used to modify the model plant *Arabidopsis* as well as crops like *Nicotiana*, maize, petunia, soybean, rapeseed, rice, apple, and fig (Zhang et al., 2018).

4.2.1.1. Modification of endogenous genes via ZFNs

Site-directed mutagenesis of a gene via ZFN-mediated system was demonstrated in *A. thaliana*. Engineered ZFNs that recognize the *ABA-insensitive 4 (ABI4)* gene were introduced into the *A. thaliana* genome via *A. tumefaciens*-mediated transformation. This gene encodes a transcription factor that regulates stress response. The ZFNs were designed to be driven by heat-shock promoter so that induction of heat shock resulted to ZFN-mediated cleavage in the *ABI4* gene. The ensuing NHEJ repair process introduced deletion and substitution in the gene in somatic cells with 3% mutation frequency. A homozygote mutant line was recovered, which exhibited ABA and glucose insensitivity. These are the expected phenotypes when *ABI4* is mutated leading to loss of its function (Osakabe et al., 2010).

Another set of ZFNs were made to recognize two other genes of *A. thaliana* namely, *alcohol dehydrogenase 1 (ADH1)* and *transparent testa 4 (TT4)*. The expression of these ZFNs were driven by estrogen-inducible promoter. They were introduced into *A. thaliana* plants by floral-dip transformation. The resulting transgenic seedlings that were induced to express the *ADH1* and *TT4* ZFNs showed 7% and 16% somatic mutation frequencies, respectively. The mutations were usually insertions of 1–2 nucleotides or deletions involving 3 to 142 nucleotides. These were all localized at the ZFN cleavage site and were likely results of NHEJ (Zhang et al., 2010). Mutagenesis of endogenous genes via ZFN was performed on the crop plant soybean. A set of ZFNs were made to target genes involved in various aspects of RNA silencing such as members of the *dicer-like* (*DCL*), *RNA-dependent RNA polymerase* (*RDR*), and *HUA enhancer1* (*HEN1*) family. Since soybean is a highly duplicated paleopolyploid plant, some genes have duplicate copies. Three of the ZFN constructs were developed to simultaneously target two paralogous gene copies namely, *DCL1a/DCL1b* and one construct targeting *DCL4a/DCL4b*. Additionally, five ZFN constructs were developed to independently target individual genes, i.e., *DCL2a, DCL2b, RDR6a, RDR6b,* and *HEN1a*. The expression of these constructs was under the control of an estrogen-inducible promoter. Evaluation of the resulting ZFN-mediated mutagenesis revealed that there were mutated alleles consisting of small insertions or deletions ranging from 1 to 20 nucleotides (Curtin et al., 2011).

4.2.1.2. Gene deletion via ZFNs

ZFN-mediated cleavage can result in the deletion of longer DNA sequences so that an entire coding region is removed (see Figure 4.5). As a proof of this concept, tobacco plants were genetically transformed with a *β-glucuronidase* (*GUS*) gene flanked by ZFN cleavage sites, a *GFP* reporter gene, and a *pat* selectable marker gene. Plants that were Basta-resistant and that showed GUS and GFP expression were selected to be crossed with another set of tobacco plants, which were genetically transformed with a ZFN gene. In several of the resulting hybrid offspring, the 4.3 kb *GUS* gene was completely deleted (Petolino et al., 2010).

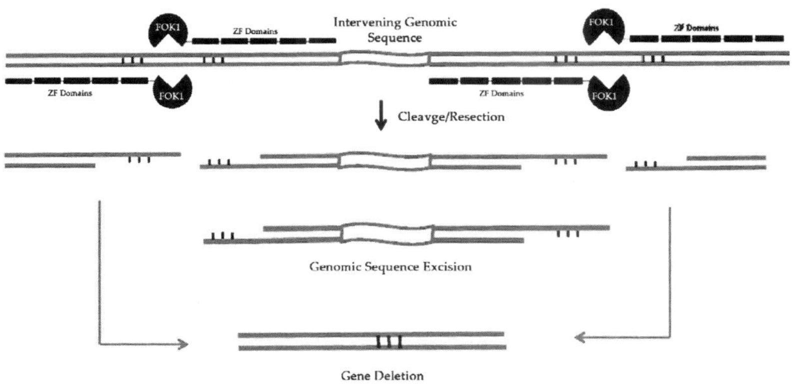

Figure 4.5: *If a gene is in between two ZFN cleavage sites, two concurrent double-strand breaks (DSBs) may take place on either side of the gene leading to its deletion.*

Source: Petolino (2015).

4.2.1.3. Gene editing by means of ZFN-mediated homology-directed repair (HDR)

It was previously discussed that DSBs on an endogenous gene may initiate an error-prone repair process, i.e., NHEJ. This may lead to mutagenesis due to the insertion or deletion of nucleotides in the gene. Another way of modifying or editing a gene is by means of HDR, requiring a donor template DNA containing the desired gene sequence modification and approximately 750 to 1,000 kb-long sequences that are homologous to those flanking the ZFN cleavage site. The DSB is repaired by the recombination of the homologous regions in the genomic cleavage site and the donor template. The desired gene sequence in the donor strand may then serve as a template so that it is copied and eventually integrated into the cleavage site (see Figure 4.4).

This was tested in *A. thaliana* which were genetically transformed with ZFNs targeting the *protoporphyrinogen oxidase* (*PPO*) gene. A donor template for the induced DSBs was also introduced, which contains a portion of the *PPO* gene and base pair substitutions that result to amino acid changes, i.e., S305L (the substitution changed amino acid at position 305 from serine to leucine) and Y426M (amino acid 426 changed from tyrosine to methionine). These two mutations render the PPO enzyme resistant to the herbicide butafenacil. Putative butafenacil-resistant plants were allowed to set seed, which were selected again on butafenacil. Eight resistant plant lines were obtained, which contain the ZFNs. Molecular analysis indicates that HDR between the genomic *PPO* locus and the donor template occurred in all lines (Pater et al., 2013).

In tobacco, specific mutations were introduced into two genes for sulfonylurea receptor, i.e., *SuRA* and *SuRB*. These mutations are known to confer resistance to imidazolinone and sulfonylurea herbicides, which inhibit the enzyme acetolactate synthase (ALS). In order to achieve this via HDR, ZFNs were created to target these *SuR* loci along with donor templates bearing each of the P191A (amino acid 191 is changed from proline to alanine), W568L (amino acid 568 changed from tryptophan to leucine), or S647T (amino acid 647 changed from serine to threonine) mutations. These were then electroporated into tobacco protoplasts. ZFN-induced herbicide resistance was observed among the calli. DNA sequence analysis revealed that out of the 12 lines generated with both the P191A and W568L donor templates, nine showed resistance due to HDR. Specifically, five and four lines showed the P191A mutation at the *SuRA* and *SuRB* loci, respectively.

Meanwhile, four, and five lines showed the W568L mutation at the *SuRA* and *SuRB* loci, respectively. On the other hand, among the 12 herbicide resistant lines with the S647T donor template, one had evidence of HDR at *SuRA* (Townsend et al., 2009).

4.2.1.4. Targeted integration of transgene via ZFNs

The typical methods for introducing a gene of interest have the downside of DNA integration into random sites in the genome. This may lead to unwanted insertional mutations as well as unpredictable effect of the transgene position on its expression. On the other hand, incorporation of the gene of interest into a specific locus significantly enhances plant transformation precision and transgenic plant performance. In addition, the stacking of transgenes into specified loci simplifies the downstream breeding process as opposed to the resource-intensive introgression performed in conventional breeding (Petolino, 2015).

4.2.1.4.1. Integration into Endogenous Loci

ZFNs can be engineered to cleave native genes and facilitate the insertion of a transgene into a native locus. A ZFN designed to cleave an *endochitinase-50 (CHN50)* gene and the transgene *pat*, an herbicide-resistance gene, flanked on each side by 750 bp of *CHN50* gene sequence were co-delivered by *A. tumefaciens* into tobacco. Up to 10% of the transgenic events were targeted incorporation of *pat* into the cleavage site at the *CHN50* locus. The integration of the transgene is mediated by HDR (Cai et al., 2009; Petolino, 2015).

A similar procedure was done in maize where targeted cleavage of the *inositol-1,3,4,5,6-pentakiphosphate kinase 1 (IPK1)* gene was facilitated by ZFNs designed to cleave exon 2 of the gene. Maize was co-transformed with these ZFNs and a donor template DNA containing a promoter less *pat* gene flanked by 815 bp that are homologous to *IPK1*. Site-specific integration of the *pat* gene at the *IPK1* locus resulted in herbicide resistance in the transgenic maize (Shukla et al., 2009; Petolino, 2015).

4.2.1.4.2. Trait Stacking by Integrating Multiple Genes

ZFN was utilized for stacking two genes for herbicide resistance in maize. The first gene, *pat*, was designed to be adjacent to a trait landing pad (TLP), which contains the recognition site of the first ZFN and sequences homologous to a second TLP adjacent to the second gene. The *pat* gene along with TLP1 was introduced into the maize genome by means of WHISKERS™-mediated transformation of embryogenic suspension cultures. Embryos that successfully integrated intact and single copy of the transgene and TLP1 were chosen for retargeting. ZFN1 along with the second gene, *aad1*, and its adjacent TLP2 were co-transformed into the selected embryos by microparticle bombardment. TLP2 contains sequences homologous to the integrated TLP 1 and the target sequence of ZFN2. ZFN1 then recognizes and cleaves its target region in TLP1 producing a DSB. This is then repaired via HDR involving TLP2 since it is homologous to the regions flanking the cleaved site. This results to *aad1* being integrated adjacent to the *pat* gene. Both herbicide resistance traits co-segregated in subsequent generations demonstrating that the two transgenes are linked or positioned adjacent to each other (Ainley et al., 2013).

4.3. TRANSCRIPTION ACTIVATOR-LIKE EFFECTOR NUCLEASES (TALENS)

Transcription activator-like effector nucleases (TALENs) are fusions of transcriptional activator-like effector (TALE) repeats and the *Fok1* restriction enzyme. The TALE proteins are derived from members of the bacterial genus *Xanthomonas*. TALE has a DNA-binding domain made up of 13–28 repeat monomers of 34 amino acids. Just like ZFNs, these repeats target specific sequence in the genome, but unlike ZFNs, each repeat recognizes a single nucleotide (see Figure 4.6). This makes them more flexible in terms of designing its target since it increases the number of potential target sites as compared to those of ZFNs (Zhang et al., 2018; Osakabe and Osakabe, 2014). This can also recognize 30 bp or longer DNA sequences giving them higher targeting specificity (Malzahn et al., 2017). After the binding of the TALE to its recognition site, the attached *Fok1* domain forms a dimer with the TALEN on the opposite strand. The dimerized *Fok1*then makes a DSB on its cleavage site. The ensuing repair process could either be NHEJ or HDR resulting to either mutagenesis, gene editing or targeted integration of a genetic sequence.

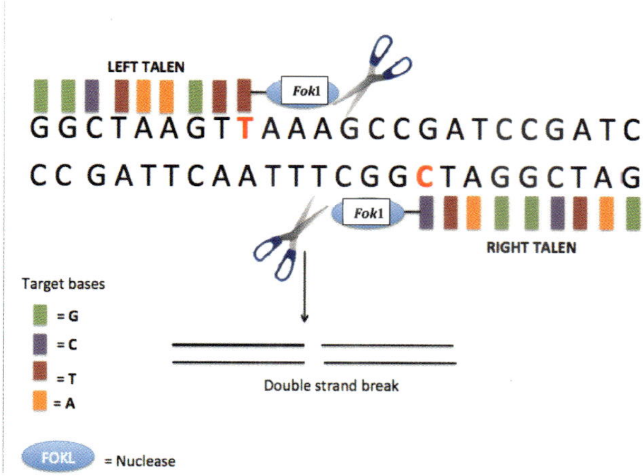

Figure 4.6: *The transcription activator-like effector nuclease (TALEN) system operates in the same manner as the zinc-finger nuclease (ZFN). Each repeat in the TALE domain recognizes and binds a specific nucleotide so that the dimerized Fok1 makes a double-strand break (DSB) at a specific cleavage site.*

Source: Ogletreerd, Creative Commons License.

4.3.1. Gene Targeting in Plants with Transcription Activator-Like Effector Nucleases (TALENs)

Targeted genome modification using TALENs has been performed in a variety of plants like *A. thaliana*, tobacco, barley, potato, tomato, sugarcane, flax, rapeseed, soybean, rice, maize, and wheat. Similar to plant genetic engineering, these endeavors to edit the genome aim to improve the crop in terms of resistance to stress, nutritional content, fragrance, and flavor (Zhang et al., 2018).

4.3.1.1. Targeted sequence deletion via TALENs

TALENs can be designed to induce the deletion of a large span of DNA segment. This was successfully done in *A. thaliana* where a pair of TALENs was designed to target and cleave two identical sequences within the duplicated gene cluster *GLL22* (see Figure 4.7). As a result, there was simultaneous cleavage of both target sites so that 4.4 kbp of intervening sequence was deleted (Christian et al., 2013).

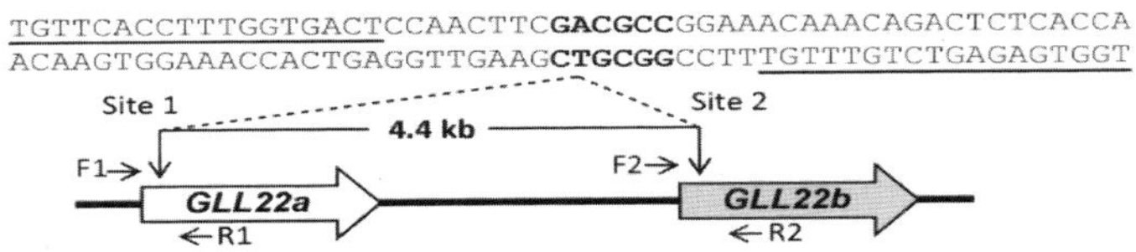

Figure 4.7: *Schematic representation of the targeted deletion of a 4.4 kbp GL22 gene cluster by a pair of TALENs (GLL22a and GL22b) that is designed to make a cut in two identical target sites with conserved sequences (Site 1 and Site 2). The conserved sequences present in both target sites are underlined.*

Source: Christian, Qi, Zhang, and Voytas (2013).

4.3.1.2. Targeted sequence modification via TALENs

Similar to ZFNs, TALENs can modify specific locus or loci by recognizing and cleaving specific sequences. The resulting DSBs initiate the NHEJ repair mechanism which introduces mutation to the target sequence.

4.3.1.2.1. Mutation of Regulatory Sequence

The promoter of the rice bacterial susceptibility gene, *Os11N3* (also called *OsSWEET14*), was targeted for TALEN-mediated disruption. This gene encodes a member of the SWEET sucrose-efflux transporter family. The bacterial pathogen *Xanthomonas oryzae* pv. *oryzae* (Xoo) has certain TALEs, i.e., AvrXa7 or PthXo3, that activate *Os11N3* expression. This results to the diversion of sugars from the plant cell to the pathogen to satisfy the pathogen's nutritional needs. The activation of *Os11N3* by the Xoo TALEs is due to the presence of effector-binding element (EBE) for AvrXa7 in the gene's promoter. This EBE overlaps with the EBE for PthXo3 and with the TATA box (Li et al., 2012).

Two pairs of TALENs were designed to induce mutation in the overlapping EBEs. This would interfere with the binding of AvrXa7 and PthXo3, thus disrupting the pathogenesis of Xoo. On the other hand, the mutation would not affect the typical expression of *Os11N3* for developmental purposes. The TALE repetitive regions included the native AvrXa7 sequence and three custom-synthesized repetitive regions. Overall, each engineered TALEN contained 24 repeat units that target a specific sequence of 24 nucleotides. Each pair of TALEN genes were cloned in a plasmid vector, which was introduced into rice embryonic cells via *A. tumefaciens*-mediated transformation (Li et al., 2012).

The successful transformants were propagated and regenerated into whole plants, constituting the T_0 generation. Then the T_0 plants were self-pollinated to produce T_1 generation. About 48% of the randomly selected T_1 plants, which were produced from the self-pollination of T_0 plants transformed with TALEN pair 1 gene, had mono or bi-allelic mutations in the *Os11N3* promoter region. On the other hand, about 63% of the randomly selected T_1 progeny carried mutations that were induced by the TALEN pair 2 genes. Most of these mutations were small deletions that did not affect the TATA box (Li et al., 2012).

The mutated *Os11N3* in T_2 plants was no longer induced by AvrXa7 or PthXo3 delivered by the Xoo bacterium. Hence, these plants also showed strong resistance to infection of the AvrXa7- or PthXo3-dependent pathogenic Xoo strain. On the other hand, all mutants were morphologically normal indicating that the developmental function of *Os11N3* was unaffected (Li et al., 2012).

4.3.1.2.2. Mutation of Endogenous Gene

Soybean lines that are low in polyunsaturated fats were produced by TALEN-induced mutations of its two *fatty acid desaturase 2* genes namely, *FAD2-1A* and *FAD2-1B*. These two genes in the seed convert the monounsaturated fat, oleic acid, to the polyunsaturated fat, linoleic acid (LA). To modify these genes, TALENs were designed to recognize and cleave conserved DNA sequences in both genes. Introduction of these to the plants resulted to mutations, specifically deletions, in the target genes by NHEJ. The resulting mutant soybeans had inactivated *FAD2-1A* and *FAD2-1B* producing nearly four times more oleic acid (from 20% to 80%) than the wild-type parents while LA drastically decreased (from 50% to 4%). This makes the resulting soybean oil healthier with longer shelf life and enhanced oxidative stability (Haun et al., 2014).

Similar experiments were performed in other crops. Targeted knockout of the *OsBADH2* gene in rice via TALEN-based NHEJ generated mutant lines with improved fragrance. This is because having a defective *Os BADH2* gene results in the production of 2-acetyl-1-pyrroline (2AP), which is a major fragrance compound in rice (Shan et al., 2015). TALEN-based mutagenesis targeted another gene in rice, *LOX3,* coding for lipoxygenase that catalyzes the dioxygenation of polyunsaturated fatty acids with at least one cis, cis-1,4-pentadiene to form hydroperoxide. LOX3-deficient rice showed enhanced seed longevity and viability (Ma et al., 2015).

Targeted mutagenesis using TALENs also proved successful in polyploid genome. An example is in bread wheat, with a hexaploid genome. Therefore, it has three homeoalleles that encode mildew-resistance locus (MLO) proteins. TALEN-induced mutation of all three MLO homeolalleles in the same wheat plant conferred heritable broad-spectrum resistance to powdery mildew (Wang et al., 2014). Another example is the autotetraploid potato. TALENs were specifically designed to knockout the *vacuolar invertase* (*VInv*) gene. This gene codes for an enzyme that breaks down sucrose into glucose and fructose. These reducing sugars accumulate in the potato tubers when these are placed in cold storage. Under high temperature (like when processing potato tubers into chips), these reducing sugars react with free amino acids resulting in brown, bitter tasting products and increased levels of acrylamide, a potential carcinogen. Mutant potato lines were recovered which had mutations in all four *VInv* alleles. The tubers from these complete knockout potatoes showed significantly lower levels of reducing sugars and therefore, the chips made from these had greatly reduced acrylamide. Furthermore, some of the mutant lines did not incorporate any TALEN DNA in their genome (Clasen et al., 2015).

4.3.1.2.3. Stacking Gene Mutations

The mutations in the *FAD2-1A* and *FAD2-1B* genes of soybeans were stacked with mutations in *fatty acid desaturase 3A* (*FAD3A*) gene in order to decrease the levels of linolenic acid, which is another type of polyunsaturated fatty acid. This would further improve the quality of the soybean oil produced. Three TALEN pairs were synthesized to recognize sequence within exon 2 and exon 3 of *FAD3A* (see Figure 4.8). The TALEN pair that exhibited the highest mutation frequency was stably integrated into the genome of the previously developed *fad2-1a fad2-1b* mutant soybean lines. Short deletions ranging from 4 to 135 bp were observed in the *FAD3A* gene. This mutation was observed to be stably inherited by the next generation (Demorest et al., 2016).

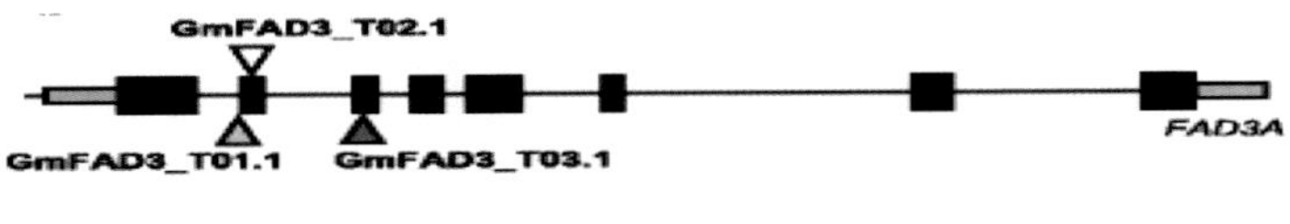

Figure 4.8: *Three pairs of TALENs were designed to recognize and cleave within the second and third exons (black boxes) of the soybean fatty acid desaturase 3A (FAD3A) gene. These are designated as GmFAD3.T01.1, GmFAD3.*

T02.1, and GmFAD3.T03.1. Among these three pairs, GmFAD3.T02.1 showed the highest mutation frequency at its target site.

Source: Demorest et al. (2016).

The seed oil from the soybean plants with triple mutations exhibited significantly lower linolenic level (2.5%) compared to the previously generated double mutant with 4.7%. In addition, LA levels further decreased from 5% to 2.7% while oleic acid levels increased from 80% to 82% (Demorest et al., 2016).

4.3.1.3. Targeted sequence insertion via TALENs

If TALENs are co-transformed with a donor template sequence into the target plant, such sequence can be inserted into the TALEN's target site via HDR. The inserted sequence can be a coding sequence containing the desired mutation, which is inserted to replace the native gene, or a promoter sequence, which is inserted to modify the gene expression level. The latter procedure was performed in tomato to modify the *anthocyanin mutant 1* (*ANT1*) gene. This gene codes for a Myb transcription factor that controls anthocyanin biosynthesis. When the strong cauliflower mosaic virus (CMV) 35S promoter was inserted upstream of the *ANT1* coding sequence (see Figure 4.9a), it leads to gene overexpression. This results to anthocyanin accumulation so that purple tissues were observed at various stages of explant regeneration (see Figure 4.9c-h). In order to achieve this, two pairs of TALENs were designed to introduce DSBs at positions ranging from 56 bp to 203 bp from the *ANT1* start codon. The donor template contained a sequence homologous to the *ANT1* promoter, the 35S promoter, and sequence homologous to *ANT1* (see Figure 4.9a). A virus vector was used to deliver both the donor template and the TALEN cassettes into the tomato cells via *A. tumefaciens*-mediated transformation (see Figure 4.9b). More than two-thirds of the insertions were precise and had no undesired sequence modifications (Čermák et al., 2015).

As early as two weeks after transformation, dark purple callus was observed on developing tomato cotyledons (see Figure 4.9c). *ANT1*-modified plants were successfully regenerated from the purple calli (see Figure 4.9d-h) producing plants with purple coloration in their flower, fruits, and leaves. Furthermore, the targeted modification was inherited by the progeny in a Mendelian fashion (Čermák et al., 2015).

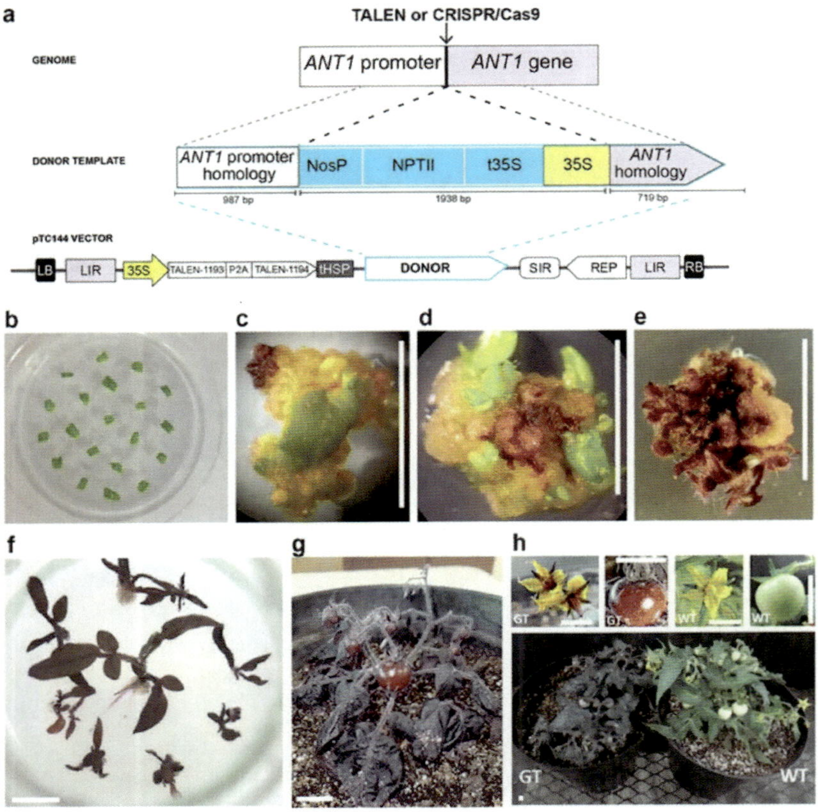

Figure 4.9: *(a) A pair of TALENs was designed to cleave between the anthocyanin mutant 1 (ANT1) promoter and the ANT1 coding sequence (top). Then via homologous recombination between the target genome sequence and the donor template (middle), the constitutive 35S cauliflower mosaic virus (CMV) promoter would be inserted upstream of the ANT1 gene. (Bottom) The ANT1 targeting or transfer DNA vector was named pTC144, which consists of the TALEN cassette and the donor template. (b) Tomato cotyledons were inoculated with A. tumefaciens containing the ANT1 vector. (c) Three weeks after inoculation, callus with the targeted insertion accumulates anthocyanin as seen in the purple area. (d) Small shoots begin to develop on the purple callus, five weeks after inoculation. (e) Multiple shoots grow from the purple callus, 10–12 weeks after inoculation. (f) Plantlets with purple leaves develop roots 12–14 weeks after inoculation. (g) Plantlet with purple fruits was transplanted into soil. (h) Dark purple coloration in flowers, fruits, and foliage were observed in genome-edited plants (GT) relative to the wild type (WT).*

Source: Čermák et al. (2015).

4.4. CLUSTERED REGULARLY INTERSPACED SHORT PALINDROMIC REPEATS (CRISPR) AND CRISPR-ASSOCIATED (CAS) PROTEINS

CRISPR systems are present in approximately 40% of bacterial species and 90% of archaean species. They were first described in *E. coli* and it was not until 30 years later that their function as an adaptive immune system of bacteria against invading nucleic acids such as plasmids or phages was exploited. This function was initially used in yogurt production or in the dairy industry, in general, to prevent phage infection in starter cultures. Eventually, the application of this system in genome modification was discovered (Metje-Sprink et al., 2019).

4.4.1. The CRISPR/Cas9 System

The classical CRISPR/Cas9 system is from *Streptococcus pyogenes* (Metje-Sprink et al., 2019), which is an RNA-directed DNA endonuclease. The CRISPR locus is made up of an operon or construct that codes for the Cas9 protein and a repeated array of spacer sequences that are transcribed and processed into CRISPR RNA (crRNA). The crRNA molecule recognizes a complementary target DNA. It anneals with a trans-activating crRNA (tracrRNA) to form a duplex. On the other hand, the crRNA and tracrRNA can be artificially fused to generate a chimeric single guide RNA (sgRNA). The crRNA-tracrRNA duplex and the sgRNA both serve as guide RNA (gRNA) that directs DNA cleavage by the Cas9 protein (see Figure 4.10) but sgRNA is simpler to use and allows for greater degree of manipulation of the CRISPR/Cas9 system. Cas9 is made up of two endonuclease domains, the HNH and RuvC-like domains that each cuts one strand. In principle, a DNA sequence that is complementary to the first 20 nucleotides of the sgRNA can be a target of Cas9. However, cleavage can only occur when a 3-nucleotide G-rich (NGG) protospacer adjacent motif (PAM) is recognized at the 3' end of the DNA targets. This means that the target sequence is made up of 23 nucleotides ending in GG (Mao et al., 2019; Osakabe and Osakabe, 2014; Malzahn et al., 2017). Similar to ZFNs and TALENs, the DSBs created by Cas9 cleavage can be repaired either via NHEJ or HDR (see Figure 4.11).

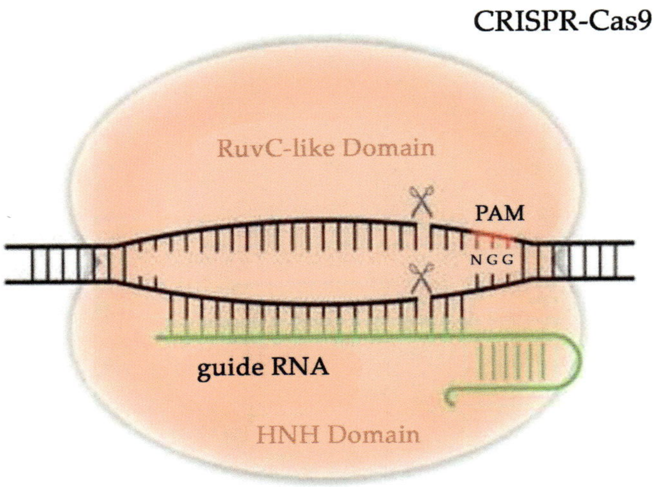

Figure 4.10: *Clustered regularly interspaced palindromic repeat (CRISPR) makes up a guide RNA (gRNA). The gRNA can either be a crRNA-tracrRNA duplex or a single guide RNA (sgRNA) formed by artificially fusing the crRNA and tracrRNA. A portion of the genome that is complementary with the first 20 nucleotides of the gRNA with a G-rich (NGG) protospacer adjacent motif (PAM) is recognized and eventually cleaved by the two endonuclease domains of Cas9, i.e., RuvC-like, and HNH domains.*

Source: Malzahn, Lowder, and Qi (2017).

The CRISPR/Cas9 construct is inserted into the T-DNA region of a binary vector. This is then introduced into the target cells via *A. tumefaciens*-mediated transformation, particle bombardment or in some cases, *in planta* transformation is performed by flower dipping. However, the latter method is not feasible in most plants. In general, promoters that induce strong expression of the CRISPR/Cas9 in meristematic cells are preferred in order to generate heritable mutations in the first-generation progeny of the transformed plants. In cases where mutant plants are regenerated from transformed cultured cells, the gene editing efficiency of the CRISPR/Cas9 system can be improved either by using strong constitutive promoters or by extending the culture period (Mao et al., 2019).

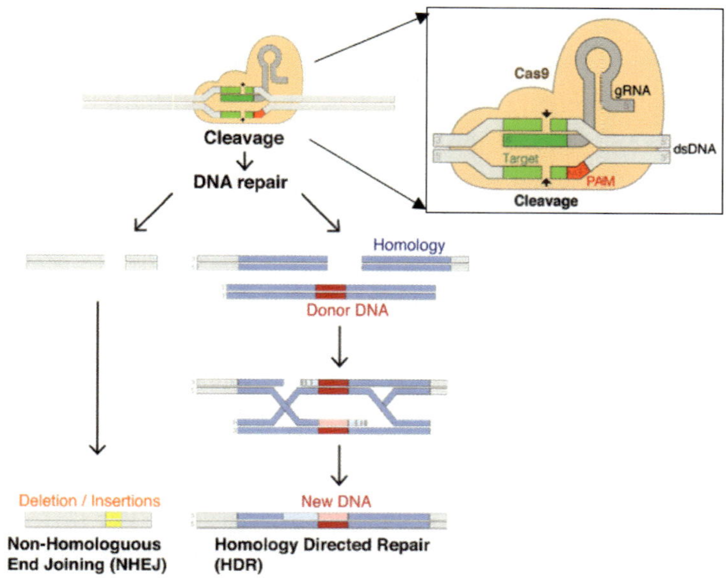

Figure 4.11: *The double-strand break (DSB) created by Cas9 cleavage can be repaired by the error prone non-homologous end-joining (NHEJ) pathway, which can introduce deletions or insertions into the target site, or homology-directed repair (HDR) pathway where genetic recombination occurs between the target endogenous segment and the donor DNA leading to the replacement of the endogenous segment by the introduced DNA.*

Source: Marius Walter, Creative Commons License.

4.4.1.1. Gene targeting in plants via CRISPR/Cas 9 system

Compared with ZFNs and TALENs, the CRISPR/Cas 9 system is simpler, more efficient, cheaper, and has the ability to target multiple genes. It is simpler because only one construct needs to be designed and synthesized to produce gRNA to target the sequence of interest. Any genomic DNA can be targeted by this system using a specifically designed gRNA together with a PAM sequence. Another advantage of this system over ZFNs and TALENs is that it can tolerate DNA methylation, which is beneficial for genome editing in plants with high GC content. A number of crops have been genetically modified by this method like rice, maize, wheat, soybean, barley, sorghum, potato, tomato, flax, rapeseed, cotton, cucumber, lettuce, grapes, grapefruit, apple, orange, and watermelon (Zhang et al., 2018; Osakabe and Osakabe, 2014; Miao et al., 2013).

4.4.1.1.1. Targeted Modification of Endogenous Genes via CRISPR-Cas9 System

The most frequent mutagenesis introduced by CRISPR-Cas9 system is small indels that result to frame shift mutations, which can introduce premature stop codons. Consequently, a targeted gene may lose its function, a process called gene knockout (Zhang et al., 2018).

4.4.1.1.1.1. Loss of Function Mutation of Endogenous Genes

Gene knockout was performed on two rice genes namely, *chlorophyll A oxygenase 1* (*CAO1*) and *LAZY1* genes. Binary T-DNA vectors for co-expression of *CAS9* and gRNA, which was either sgRNA or crRNA-tracrRNA duplex, were transformed into rice. The gRNA was designed to target either the second exon of the *CAO1* gene or the third exon of the *LAZY1* gene. The sgRNA targeted the genes more efficiently than the crRNA-tracrRNA duplex. Insertions or deletions in the *CAO1* gene were observed 3–4 bp upstream of PAM, which is consistent with the location of Cas9 cleavage site. Such mutation is expected to make the gene non-functional so that chlorophyll synthesis becomes defective. Lines that showed biallelic mutation of *CAO1* had greatly reduced chlorophyll B (Chl B) content, which would explain the pale green leaves of these lines. Likewise, the *LAZY1* gene was also disrupted resulting to a pronounced tiller-spreading phenotype (Miao et al., 2013).

Another set of four genes in rice, which are regulators of certain yield traits, were also targeted for mutagenesis. These are *grain number 1a* (*Gn1a*), *dense and erect panicle 1* (*DEP1*), *grain size 3* (*GS3*), and *ideal plant architecture 1* (*IPA1*). *Gn1a* codes for a cytokinin dehydrogenase 2 which regulates the number of grains per panicle. *DEP1* and *GS3* both encode the γ subunit of G protein but vary in their function. *DEP1* specifies an erect panicle and influences plant height as well as grain size while *GS3* regulates grain and seed size. Finally, *IPA1*, also known as *OsSPL14*, specifies a squamosa promoter binding protein which regulates plant height and tiller number (Li et al., 2016).

The sg-RNA-Cas9 expression vectors that target each gene were introduced into *A. tumefaciens* by electroporation. Then *A. tumefaciens*-mediated transformation of rice embryonic calli was performed. The CRISPR/Cas9 construct for *Gn1a* was specifically designed to introduce mutation in the first exon, which would modify the entire coding region of the gene. Sequencing of the targeted region in the resulting mutant lines revealed insertion and deletion of different nucleotides. The mutant lines were grown in the field and homozygous T2 plants showed increased plant height, panicle size, and number of flowers per panicle as compared to the wild type (see Figure 4.12) (Li et al., 2016).

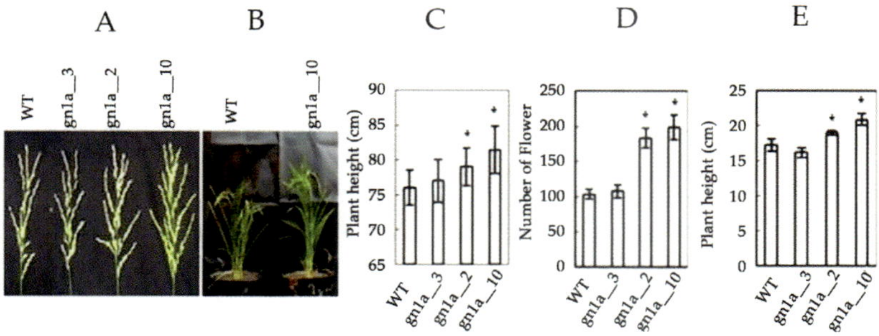

Figure 4.12: *(A,E) The Gn1a knockout mutants (gn1a—2 with a 2-bp deletion and gn1a—10 with a 10 bp deletion) showed longer panicles than the wild type. (B,C) Their plant height is also significantly higher, and (D) they have more average numbers of flowers per panicle than the wild type. The other mutant line with a 3-bp deletion (gn1a—3), which resulted to a single amino acid deletion, showed similar phenotype in terms of panicle size, plant height, and flower number relative to the wild type.*

Source: Li et al. (2016).

Another CRISPR/Cas9 construct was engineered to introduce a mutation in exon 5 of *DEP1*. Homozygous T2 mutants were grown in the field and two of them with frame-shift mutations, *dep1–2* (with a 2-bp deletion) and *dep1–4* (with a 4-bp deletion), had decreased plant height and short panicles relative to the wild type (see Figure 4.13). On the other hand, the mutant lines showed an increased number of flowers per panicle indicating that mutation of *DEP1* can result to dense panicles. Mutant lines also had significantly smaller grain size than the wild type (Li et al., 2016).

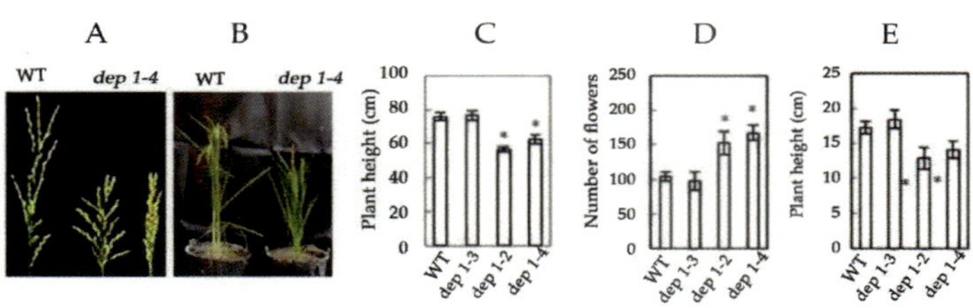

Figure 4.13: *DEP1 knockout mutants (dep1–4 with 4-bp deletion and dep1–2 with 2-bp deletion)had significantly shorter panicles (A,E) and plant height (B,*

C), but showed more flowers per panicle (D) compared to the wild type. Another mutant line, dep1–3 with 3-bp deletion, served as a control which exhibited the same phenotype as the wild type.

Source: Li et al. (2016).

The third type of CRISPR/Cas9 construct contained sgRNAs that aim to introduce deletion at a specific region of the *GS3* gene, i.e., 165 bp downstream from the predicted start codon. Homozygous T2 frameshift mutants with a 4-bp deletion (*gs3–4*) and with a 5-bp deletion (*gs3–5*) exhibited larger grain size and long awns on the husks (see Figure 4.14). The significantly longer grains resulted to increased grain weight. The number of flowers per panicle did not show any significant difference between the mutant lines and the wild type. This indicates that *GS3* is involved in regulating grain size and its mutation can increase grain size (Li et al., 2016).

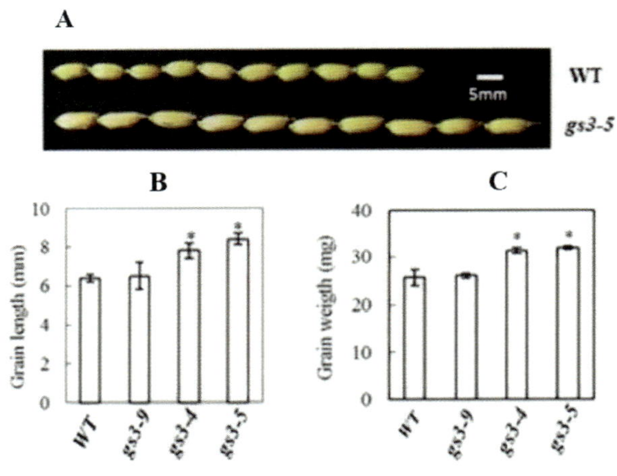

Figure 4.14: *(A,B) The GS3 knockout mutants with 4-bp deletion, gs3-4, and with 5-bp deletion, gs3-5, showed significantly longer grains compared to the wild type. This also resulted to higher grain weight in these mutants (C). The mutant with 9-bp deletion, gs3-9, had three amino acids deleted and served as the control with phenotypes like the wild type.*

Source: Li et al. (2016).

Finally, the *IPA1* knockout mutants were generated by designing a CRISPR/Cas9 construct that targets bases 854 to 876 from the start codon of the cDNA. This region is close to the site where OsmiR156c leaves *IPA1* transcripts. There are varying phenotypes observed depending on the type of mutation produced. The *ipal–3* mutant with a 3-bp deletion did not induce any frame shift mutation in the coding region and did not affect the OsmiR156 target site so that it exhibited phenotype similar to the wild type (see Figure 4.15). Two mutants, *ipal–12* and *ipal–21*, had deletion of 12 bp and 21 bp, respectively. This resulted to deletions of certain amino acids but no shift in the reading frame. The IPA1 protein activity was maintained but the OsmiR156 target site in the IPA1 mRNA was changed. These mutants showed much fewer tillers compared to the wild type, but their plant height, flower number, and panicle length are all increased. The third class of mutant, *ipal–5*, had a 5-bp deletion which caused a frame shift resulting in a non-functional protein product. They showed a dwarf phenotype with increased tiller number (Li et al., 2016).

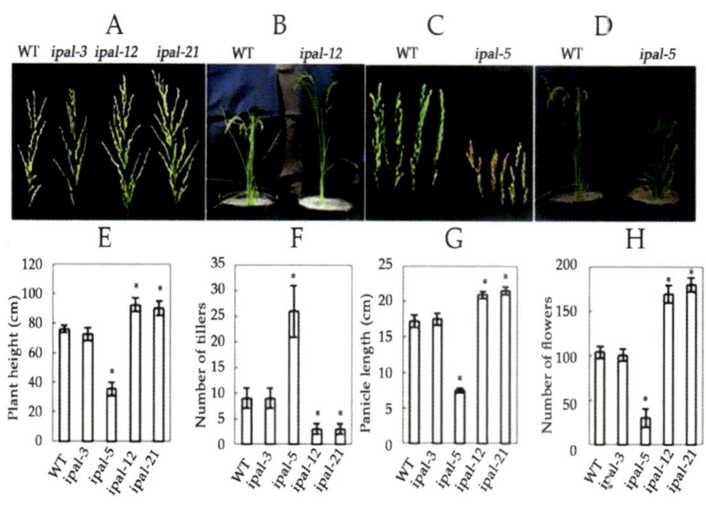

Figure 4.15: *The IPA1 knock out mutants with 12-bp and 21-bp deletion, ipal–12 and ipal–21, respectively, showed longer panicles (A,G), increased plant height (B, E) and more flowers (H) but lesser tiller number (F) compared with the wild type. Another mutant, ipal–5, with a 5-bp deletion exhibited significantly reduced panicle length (C, G), a dwarf phenotype (D, E), lesser flowers (H) but greatly increased number of tillers (F) compared to the wild type. The mutant with 3-bp deletion, ipal–3, showed wild type phenotypes (A, E-H).*

Source: Li et al. (2016).

The nutritional content of crops can also be enhanced by mutating certain genes using CRISPR/Cas9. This was done to improve the oleic acid content of the oil seed plant *Camelina sativa* as well as to increase the amylopectin proportion in rice and maize. Furthermore, plant resistance to bacterial, viral, or fungal diseases was improved by mutagenesis via the CRISPR/Cas9 technology. This was done in rice, tomato, citrus plants, and cucumber (Zhang et al., 2018).

4.4.1.1.1.2. Targeted Mutagenesis via Transient Expression of CRISPR/Cas9 DNA or RNA

The introduced CRISPR/Cas9 DNA is usually integrated into the host plant chromosome, which increases the chance of off-target changes. This also places the resulting mutant lines in the same category as genetically modified organisms, which causes legislation difficulties. This led to the development of transgene-free methods of genome editing (Zhang et al., 2016) (Figure 4.16).

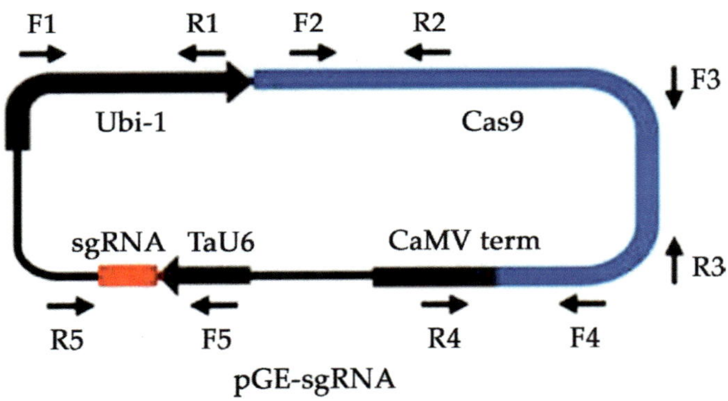

Figure 4.16: *Schematic of the structure of the transiently expressed CRISPR/Cas9 vector containing the sgRNA expression cassette and the Cas9 coding sequence. Five primer sets were designed to detect if the CRISPR/Cas9 DNA was integrated into the genome of the target plants.*

Source: Zhang et al. (2017).

One such approach involves the transient expression of CRISPR/Cas9 DNA. This was performed on both bread wheat, which is a hexaploid, and durum wheat, which is a tetraploid. In the case of bread wheat, a sgRNA that targets the highly conserved exon 3 of the three homeologues of *TaGASR7* (i.e., *TaGASR7-A1, TaGASR7-B1,* and *TaGASR7-D1*) was designed. These genes have been implicated in regulating grain length and weight. The sgRNA expression cassette was combined with that of Cas9 in a single DNA construct (see Figure 4.16). This was introduced into immature embryos of two varieties of bread wheat (i.e., Bobwhite, and Kenong 199) via particle bombardment (see Figure 4.17). After transient expression and function, the CRISPR/Cas9 DNA was degraded. Seedlings were regenerated from the bombarded embryos. In contrast with most plant genome-editing experiments, herbicide or antibiotics were not added to the culture medium to select for transformed plants. Among the 80 Bobwhite mutants analyzed, 51 showed targeted indels in all the three *TaGASR7* homeologues. Out of these 51 mutants, eight had all six alleles simultaneously knocked out (Zhang et al., 2016).

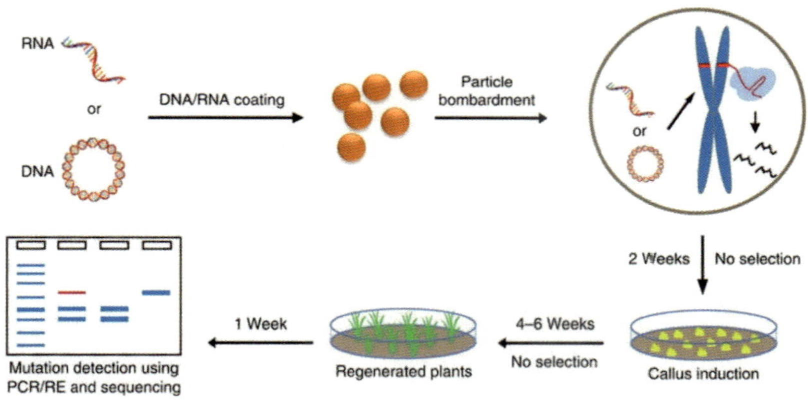

Figure 4.17: *Overview of the genome-editing methods using transiently expressed CRISPR/Cas9 DNA or RNA. The designed constructs were introduced into immature embryos via particle bombardment and then the calli were regenerated in medium without herbicide selection. The presence of targeted mutation is determined via PCR-restriction enzyme digestion (PCR/RE) assay.*

Source: Zhang et al. (2017).

Mutation was also introduced into the two *TdGASR7* homeologues (i.e., *TdGASR7-A1* and *TdGASR7-B1*) in durum wheat via transiently expressed CRISPR/Cas9 DNA. Homozygous mutants with all four alleles modified were obtained from the seedlings that were regenerated after particle bombardment (Zhang et al., 2016).

Since the mutant seedlings were regenerated in the absence of herbicide selection, there is a high probability that the CRISPR/Cas9 DNA construct was not integrated into the wheat genome. PCR analysis determined that the construct is absent in 43.8% of the T0 Bobwhite mutants and in 61.9%

of the Kenong199 T0 mutants. Likewise, 54.5–58.3% of the T0 mutants of the two durum wheat varieties showed absence of the CRISPR/Cas9 DNA construct. This shows that plants with targeted mutations and without active transgenes can be obtained through this genome-editing procedure (Zhang et al., 2016).

However, a certain percentage of the mutants obtained still carried CRISPR/Cas9 DNA. Also, there is a possibility that the degradation products of CRISPR/Cas9 DNA maybe integrated into the genome. Thus, this procedure was further improved by transiently expressing *in vitro* transcripts of Cas9-coding sequence and sgRNA in wheat callus cells. The transiently expressed CRISPR/Cas9 RNA was designed to introduce mutations into the conserved region in exon 8 of *TaGW2* gene, which functions as a negative regulator of kernel width and weight in bread wheat. The RNAs were delivered into the immature embryos of the Kenong199 variety by particle bombardment (see Figure 4.16). Regeneration of the embryos was done without herbicide selection. From the 1,600 bombarded embryos, 17 T0 mutants were identified (i.e., 1.1% efficiency), among which six mutants contained indels in all six *TaGW2* alleles. Since RNA molecules are less stable than DNA and are easily degraded, it is unlikely to be integrated into the nuclear genome. However, the instability of RNA may also explain the lower mutagenesis frequency of this method relative to using CRISPR/Cas9 DNA (Zhang et al., 2016).

Another strategy to reduce transgene integration is by directly introducing assembled Cas9 ribonucleoproteins (RNPs) into regenerative cells. This was performed in *A. thaliana*, lettuce, tobacco, and rice where Cas9 RNPs were transfected into protoplasts followed by tissue regeneration (Woo et al., 2015). Cas9 RNP is made up of Cas9 protein and gRNA. It is degraded after delivery, which also reduces off-target effects (Merck, 2019).

4.4.1.1.2. Genome Editing by Sequence Insertion and Replacement via CRISPR/Cas9 System

Several traits that are agriculturally significant are enhanced not by knocking out a certain gene but by changing the gene expression level. This can be accomplished by point mutation or base pair substitution which is not typically produced by NHEJ. Therefore, HDR is utilized in order to achieve gene replacement where a sequence with the desired base substitution can be inserted into a target location. On the other hand, the frequency of gene integration by HDR is lower than the mutation frequency effected by NHEJ

(Sun et al., 2016). In other cases, the regulatory elements, instead of the coding sequence, are targeted for editing.

4.4.1.1.2.1. Gene Editing by CRISPR/Cas9 System

A system of CRISPR/Cas9-mediated gene insertion was employed to confer herbicide resistance in crops like rice, soybean, maize, and potato. In this system, a donor repair template contains a genetic sequence that is desired to be expressed instead of the endogenous gene.

In rice, multiple point mutations were introduced into its *Acetolactate synthase 1* (*ALS1*) gene using CRISPR/Cas9-mediated homologous recombination. This gene codes for the ALS, which catalyzes the first step in the biosynthesis of branched amino acids (i.e., leucine, isoleucine, valine). It is a major target for at least four classes of herbicides namely, sulfonylureas, imidazolinones, triazolopyrimidine sulfonamides, and pyrimidinyl carboxy groups (Sun et al., 2016; Butt et al., 2017). A plasmid vector was designed to contain the CRISPR/Cas9 expression cassette as well as two gRNAs. Having two instead of the usual single gRNA increases the efficiency of HDR. The plasmid also contained the donor template for HDR, which had a portion of the *ALS* gene but with the desired substitutions, i.e., W548L (amino acid 548 is changed from tryptophan to leucine) and S627I (amino acid 627 is changed from serine to isoleucine). The donor template also contained base substitutions that would not change the amino acid residues but would prevent the Cas9 from cleaving the donor fragment again once HDR occurred. Both ends of the donor template were target sequences of the two gRNAs including the PAM motifs. These would enable the donor DNA to be released via Cas9 cleavage of the targets (Sun et al., 2016).

To ensure that there would be sufficient donor templates and thus, increase HDR efficiency, free 477-bp donor fragment, and the plasmid vector were simultaneously introduced into rice calli via particle bombardment. Most of the generated plants underwent HDR so that the target bases showed the W548L and S627I substitutions. Both wild type and mutant lines were sprayed with bispyribac sodium (BS) herbicide at the five-leaf stage. BS belongs to the pyrimidinyl carboxyherbicide group. The leaves of the wild type withered after ten days and the plant eventually died after 36 days. In contrast, the mutant lines grew normally exhibiting herbicide tolerance (Sun et al., 2016).

A similar experiment targeting the same gene in rice was performed but using a chimeric single-guide RNA (cgRNA). This construct contained both the sgRNA and the donor repair template. Several types of cgRNA were designed and utilized to evaluate for gene editing efficiency. The cgRNA functions as a *cis* repair template if the sgRNA and the repair template work as a single bifunctional molecule (see Figure 4.18A). In this scheme, Cas9 functions both as an endonuclease that produces DSB and as a loading scaffold that mediate the targeting of the RNA repair template to the DSB site. Meanwhile, the *trans* repair template had the sgRNA and the repair template separated by a tRNA and this was further processed to split the sgRNA from the repair template (see Figure 4.18B). The repair templates were also designed to vary in terms of orientation. Those with antisense orientation complement the non-target strand while those with sense orientation complement the target strand (see Figure 4.18A and B) (Butt et al., 2017).

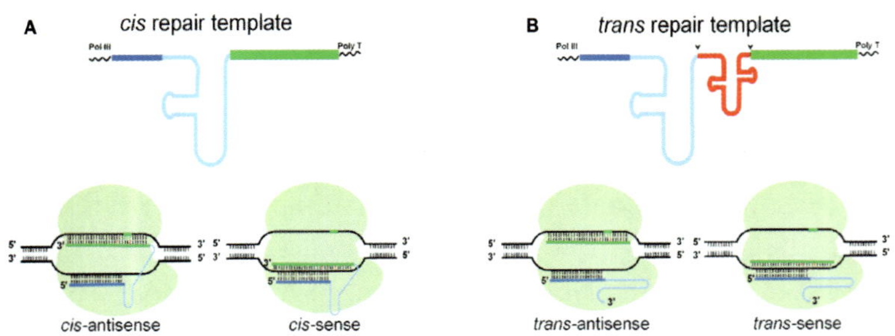

Figure 4.18: *There are several types of chimeric single guide RNAs (cgRNA) based on modality of function and orientation. (A) The cis repair template has the guide RNA (gRNA) (blue) attached directly to the repair template (green). The repair template is in antisense orientation if it is complementary to the non-target strand. It is in the sense orientation if, just like the gRNA, it is complementary to the target strand. (B) The trans repair template, has a tRNA (red) separating the gRNA from the repair template. It is cleaved at certain sites (indicated by arrows) to separate the two RNA molecules. Similar to the cis type, the trans repair template can either have an antisense or sense orientation relative to the gRNA. (Light blue line represents the scaffold RNA, green vertical lines represent the PAM sequence, red vertical lines represent the introduced mutations).*

Source: Butt et al. (2017).

Rice protoplasts were transfected with expression cassettes carrying Cas9 and cgRNA that were designed to target *ALS1*. All types of cgRNA were able to generate DSBs but the repair efficiency was highest in the *cis-*

antisense repair templates. In order to generate tolerance to the herbicide BS, the different types of cgRNA were designed to introduce a segment of the *ALS1* gene with the W548L point mutation. Specifically, this involves changing the TGG codon to TTG. Rice calli were genetically transformed with these constructs via *A. tumefaciens*. Among the mutants recovered, regeneration was observed in plantlets with the *cis*-antisense template. Gene editing and herbicide tolerance were observed with *cis*-antisense repair templates, specifically those using the human scaffold RNA sequence (see Figure 4.19) (Butt et al., 2017).

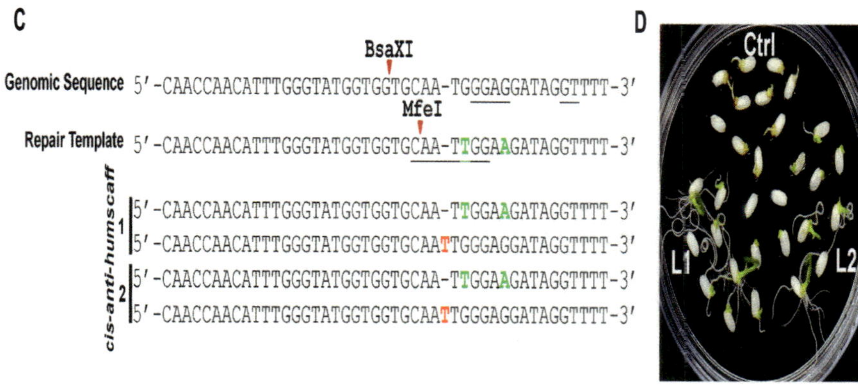

Figure 4.19: *(A) Mutant rice lines with the cis-antisense cgRNA template with human scaffold RNA showed the desired point mutations (green nucleotides) in the acetolactate synthase 1 (ALS1) gene indicating a successful homologous recombination between the target sequence in the genome and the repair template. Some insertions were also observed (red nucleotide). (B) Seeds from these mutant lines (L1 and L2) germinated on culture with the herbicide bispyribac sodium (BS) showing tolerance to this herbicide as opposed to the wild type (Ctrl).*

Source: Butt et al. (2017).

4.4.1.1.2.2. Editing Regulatory Sequences by CRISPR/Cas9 System

In maize, this method was used to increase the expression level of the *ARGOS8* gene by editing the promoter region. This was done by inserting a GOS2 promoter (GOS2 PRO), which induces moderate-level expression, into the 5'-untranslated region (5'-UTR) of the gene or using it to replace the native promoter of the gene (see Figure 4.20) (Shi et al., 2016).

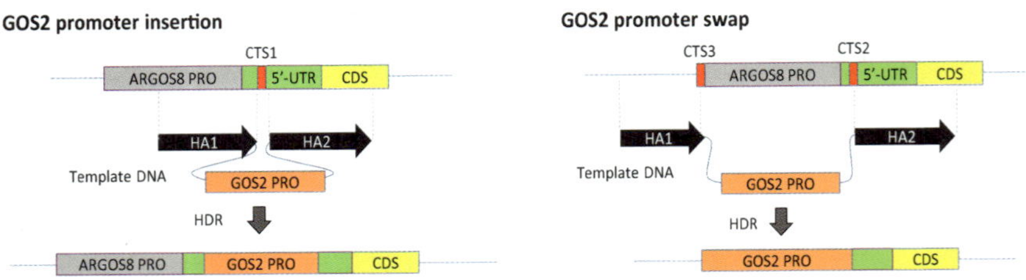

Figure 4.20: *CRISPR/Cas9-mediated gene editing was used to insert GOS2 promoter (GOS2 PRO) into the 5'-untranslated region (5'-UTR) of the ARGOS8 gene. The gRNA was designed to guide the Cas9 to make a cut at the CRISPR-RNA target site 1 (CTS1). Homology-directed repair (HDR) of the cleavage using the template DNA containing the GOS2 PRO flanked by homology arms (HA1 and HA2) resulted to insertion of the desired promoter. This method was also used to replace entirely the native promoter of ARGOS8 (ARGOS8 PRO) with the GOS2 PRO. The main difference with the previous method was the Cas9-mediated cleavage of target sites (CTS2 and CTS3) flanking ARGOS8 PRO in order to remove it from the genome. GOS2 PRO is expected to increase the expression level of the ARGOS8 coding sequence (CDS).*

Source: Shi et al. (2016).

Precise genome editing at the *ARGOS8* promoter locus was confirmed by PCR and sequencing. The mutant lines had elevated levels of *ARGOS8* mRNA transcripts relative to the wild type. *ARGOS8* codes for a negative regulator of ethylene response so that its overexpression is expected to reduce ethylene sensitivity and improve grain yield under drought stress conditions. A field study showed that compared to the wild type, the mutant lines increased grain yield by five bushels per acre under drought conditions that occurred during flowering stage (Shi et al., 2016).

CRISPR-mediated changes in the promoter of tomatoes was also done to enhance overall yield. Promoters of genes influencing plant architecture, shape, and fruit size of tomato were manipulated. For instance, the expression of genes affecting the number of floral organs and locules was altered in order to enhance fruit size (Coleman, 2017).

4.4.1.1.2.3. Editing of Multiple Sequences via CRISPR/Cas9 System

The CRISPR system can be utilized for trait stacking and can also be designed to simultaneously target multiple genes in a single cell. This process is called multiplexing. If the genes are homologous, then only one or two sgRNAs are engineered for target recognition. However, if the target genes have no sequence similarity, then sgRNAs will be required for each target. This means that multiple RNAs are co-expressed within a single construct and introduced into a single binary vector (Mao et al., 2019).

One way to achieve this is by designing a two-component transcriptional system where the expression of Cas9 and gRNAs or sgRNAs are driven by separate transcriptional regulatory units. Strong promoters are usually utilized in order to induce maximal expression of sgRNAs and Cas9 in plants. For regulating Cas9 expression, RNA polymerase II (RNA pol II)-dependent promoters are used, which drive strong expression in reproductive cells. On the other hand, RNA polymerase III (RNA pol III)-dependent promoters are more suited for transcriptional control of sgRNAs (see Figure 4.21) (Mao et al., 2019). Each of the multiple sgRNAs can have its separate RNA Pol III-dependent promoter to enable efficient co-expression (see Figure 4.21 Ia). This was done to simultaneously target the mutation of up to eight genes in rice and *A. thaliana* (Mao et al., 2019; Ma et al., 2015). This strategy can be simplified by assembling the multiple sgRNAs into a single transcriptional unit driven by a single promoter (see Figure 4.21Ib). Self-cleaving RNAs such as ribozyme and tRNA sequences are used to process the primary transcript to generate multiple mature sgRNAs. This was used to simultaneously target four homologous genes in rice, i.e., *MAPK1, MAPK2, MAPK5,* and *MAPK6*, by using eight sgRNAs. The endogenous tRNA-processing system was utilized to process the polycistronic gene and produce numerous sgRNAs (Mao et al., 2019; Xie et al., 2015). The addition of a 3' synthetic polyadenine (polyA) sequence can promote further polyadenylation and mRNA export from the nucleus. Thus, it is placed at the end of the Cas9 coding sequence to ensure its translation (Tang et al., 2016).

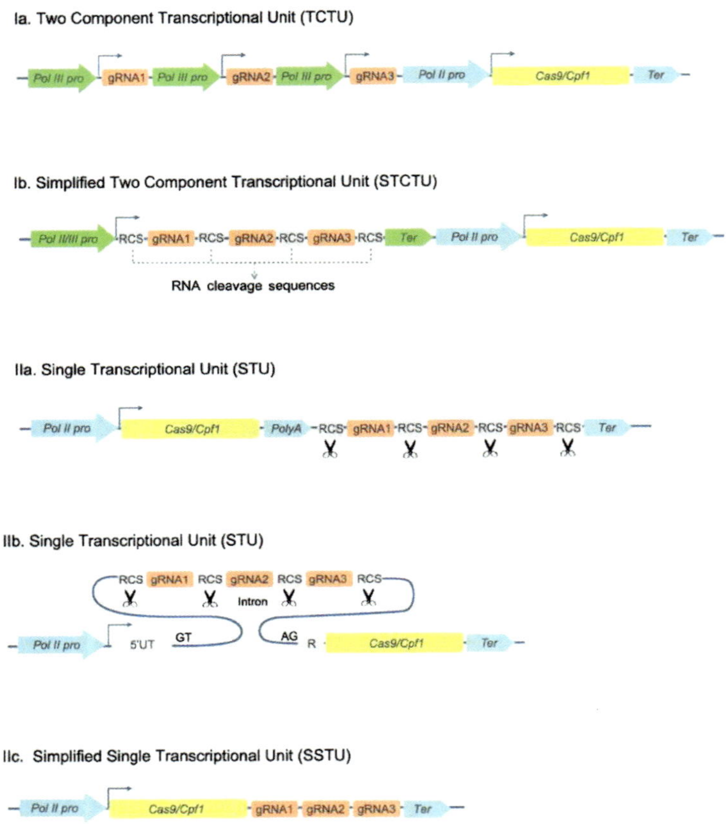

Figure 4.21: *Multiplex gene editing via the CRISPR/Cas9 system can be accomplished by designing a polycistronic unit containing the codes for several gRNAs, Cas9, and/or RNA cleavage sequences (RCS). The stronger RNA polymerase II-dependent (Pol II) promoter is usually used to direct Cas9 expression while gRNA expression is driven by either a Pol II or RNA polymerase III-dependent (Pol III) promoter. The gRNAs can either (Ia) have their individual promoters or have (Ib) a single promoter to direct the expression of all the gRNAs in the unit. (IIa) A single promoter can also be used to control the expression of Cas9 and the gRNAs. (IIb) Another version of this strategy is to engineer the code for gRNA and RCS into the introns. (IIc) The single transcriptional unit can be simplified further by removing the RCS. (PolyA: polyadenylation sequences; 5'UT-R: 5' untranslated region).*

Source: Mao, Botella, Liu, and Zhu (2019).

Another strategy for simultaneous targeting of multiple genes is by designing a single transcriptional unit. This time, a single strong promoter regulates the expression of Cas9 and the numerous sgRNAs (see Figure 4.21 IIa). This allows the synchronization of the expression patterns of Cas9 and sgRNA. This was performed on rice where Cas9, two sgRNAs and a self-cleaving ribozyme were co-expressed from a single RNA Pol II-dependent promoter in order to target *OsPDS* gene (Tang et al., 2016). To ensure the expression of the construct in a specific plant cell or developmental stage, inducible, or tissue- and development-specific promoters can be utilized. A variation of this method is to construct sgRNAs and self-cleaving RNA molecules into the introns or untranslated regions (UTR) of the transcriptional unit (see Figure 4.21 IIb). A simpler single transcript can be designed without any self-cleaving RNA sequences (see Figure 4.21 IIc). The primary transcript is processed by endogenous RNA processing proteins in order to release mature sgRNAs (Mao et al., 2019).

4.4.1.2. Improving CRISPR/Cas9 specificity

Specific target recognition by genome engineering tools are critical since off-target cleavage results to unwanted and unexpected mutation that has generally detrimental effects. As discussed previously, transient expression of CRISPR/Cas9 system reduces off-target effects. This can also be avoided by increasing the system's specificity. It is previously described that the specificity of CRISPR/Cas9 in recognizing the target is determined by a 20-nucelotide gRNA. Generally speaking, the degree of complementarity between the gRNA and the target DNA determines target recognition. However, recent studies suggest that the 3'end of the target DNA that flank PAM is more restrictive when pairing with the gRNA, but mismatches are tolerated at the 5' end (Osakabe and Osakabe, 2010).

Certain strategies were used to further improve the target specificity of the CRISPR/Cas9 system. One of these is using a mutated version of Cas9 called Cas9 nickase (Cas9n) with one of the two endonuclease domains inactivated. This means that Cas9n can induce nicks or single-stranded DNA breaks (SSBs) (see Figure 4.22a). Two sets of Cas9n can then be used with distinct gRNAs in order to generate a DSB. The use of a pair of gRNAs increases the length of the target sequence from 20 nucleotides to 40 nucleotides. This strategy could reduce off-target effects by 20–1,500-fold. In *A. thaliana*, the use of nickases induced HDR but failed to induce NHEJ, confirming that SSBs are repaired with high fidelity (Osakabe and Osakabe, 2010; Malzahn et al., 2017; Fauser et al., 2014).

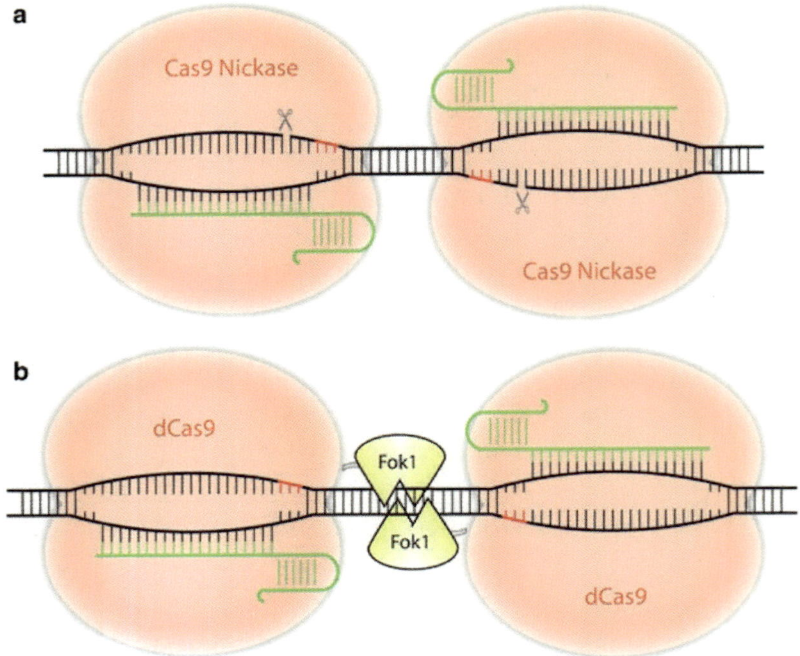

Figure 4.22: *(a) Cas9 can be modified to become a nickase, inducing single-stranded breaks in the target site. Therefore, a pair of nickases is required to generate a double-stranded break (DSB). (b) Cas9 may also be inactivated or rendered catalytically dead (dCas9) and then fused with Fok1 nuclease, which generates the DSB when it dimerizes with another Fok1 nuclease at the target site.*

Source: Malzahn, Lowder, and Qi (2017).

Another way of increasing CRISPR target specificity is by fusing a catalytically inactive or catalytically dead Cas9 (dCas9) with a Fok1 nuclease domain (see Figure 4.22b). Since Fok1 nuclease requires dimerization to be functional, a pair of gRNAs and Fok1-dCas9 is required. The gRNAs lead the dCas9 to the target sequence and then the Fok1 nuclease domains dimerize to cleave the DNA. As with the paired nickase strategy, the requirement of two gRNAs in this method reduces off-target effects (Malzahn et al., 2017).

4.4.2. Other CRISPR Systems

Apart from the CRISPR/Cas9 system, there are other Cas variants from other species like *Staphylococcus aureus* and *Streptococcus thermophilus*. Cas9 derived from *S. aureus* and *S. thermophilus* were shown to work efficiently in plants (Mao et al., 2019). Cas12a, formerly called Cpf1, from *Prevotella* and *Francisella* can also be used. The CRISPR/Cas12a system differs from the classical system in terms of having smaller nucleases, having a naturally occurring sgRNA, and the cut done by Cas12a results in staggered ends as opposed to the blunt ends produced by Cas9 cleavage. In addition, the PAM in the Cas12a system has to be rich in thymine instead of guanine, which can diversify the potential target sites for genome editing. Cas12a makes the cut distal from the recognition site while Cas9 cleaves proximal from the site (Mao et al., 2019). This has been successful in performing targeted mutagenesis in tobacco and rice, in inducing targeted gene insertions in rice, and in performing transgene-free genome editing also in rice (Zhang et al., 2018).

Another variant, Cas13a, is able to recognize and cleave RNA instead of DNA (see Figure 4.23). Hence, this is used for RNA editing in bacteria, mammals, and plants. The RNase activity of Cas13a is due to its two higher eukaryotes and prokaryotes nucleotide-binding domains (HEPN). A CRISPR/Cas13a system was employed to provide molecular immunity to *Nicotiana benthamina* against the Turnip mosaic virus (TuMV), an RNA virus. A recombinant TuMV expressing green fluorescent protein (TuMV-GFP) was used to allow the monitoring of virus infection and replication. The Cas13a expression construct was optimized to be expressed in plants by synthesizing four overlapping fragments, which were flanked by attachment site L1 (attL1) and attL2 recombination sites and a nuclear localization signal (NLS) fused to the C-terminus of the protein. This construct was cloned into a binary vector along with CaMV 35S promoter, which induces overexpression. The vector was then introduced into *A. tumefaciens* via electroporation. In order to target the TuMV-GFP genome, crRNAs complementary to four different regions of the TuMV-genome were designed and introduced into *A. tumefaciens.* Then *N. benthamina* were agroinfected with mixed *A. tumefaciens* carrying Cas13a and engineered RNA2 genome containing crRNAs against one of the four different TuMV-GFP genome targets. Results indicate that effectiveness of the CRISPR/Cas13a system in attenuating TuMV replication and spread (Aman et al., 2018; Mao et al., 2019).

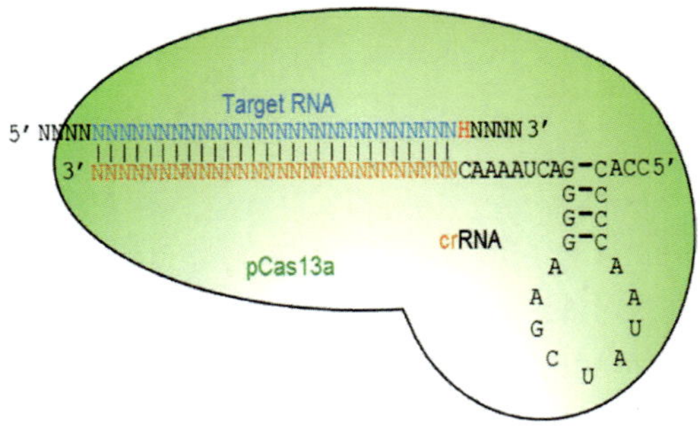

Figure 4.23: *The CRISPR/Cas13a system is composed of a crRNA that recognizes and binds the target RNA, which is then degraded by Cas13a.*

Source: Aman et al. (2018).

REFERENCES

1. Ainley, W. M., Sastry-Dent, L., Welter, M. E., Murray, M. G., Zeitler, B., Amora, R., et al., (2013). Trait stacking via targeted genome editing. *Plant Biotechnology Journal*, *11*(9), 1126–1134. https://doi.org/10.1111/pbi.12107 (Accessed on 18 December 2019).
2. Aman, R., Ali, Z., Butt, H., Mahas, A., Aljedaani, F., Khan, M. Z., Ding, S., & Mahfouz, M., (2018). RNA virus interference via CRISPR/Cas13a system in plants. Commons attribution 4.0 international license. *Genome Biology*, *19*, 1. https://doi.org/10.1186/s13059-017-1381-1 (Accessed on 18 December 2019).
3. Butt, H., Eid, A., Ali, Z., Atia, M. A. M., Mokhtar, M. M., Hassan, N., Lee, C. M., Bao, G., & Mahfouz, M. M., (2017). Efficient CRISPR/Cas9-mediated genome editing using a chimeric single-guide RNA molecule. Creative commons attribution license. *Frontiers in Plant Science*, *8*, 1441. doi: 10.3389/fpls.2017.01441.
4. Cai, C. Q., Doyon, Y., Ainley, W. M., Miller, J. C., DeKelver, R. C., Moehle, E. A., et al., (2009). Targeted transgene integration in plant cells using designed zinc finger nucleases. *Plant Molecular Biology*, *69*(6), 699–709. https://doi.org/10.1007/s11103-008-9449-7 (Accessed on 18 December 2019).
5. Carroll, D., (2011). Genome engineering with zinc-finger nucleases. *Genetics*, *188*(4), 773–782. doi: 10.1534/genetics.111.131433.
6. Čermák, T., Baltes, N. J., Čegan, R., Zhang, Y., & Voytas, D. F., (2015). High-frequency, precise modification of the tomato genome. Creative commons attribution 4.0 international license. *Genome Biology*, *16*, 232. doi: 10.1186/s13059-015-0796-9.
7. Christian, M., Qi, Y., Zhang, Y., & Voytas, D. F., (2013). Targeted mutagenesis of *Arabidopsis thaliana* using engineered TAL effector nucleases. Creative commons 3.0 unported license. *G3 (Bethesda)*, *3*(10), 1697–1705. doi: 10.1534/g3.113.007104.
8. Clasen, B. M., Stoddard, T. J., Luo, S., Demorest, Z. L., Li, J., Cedrone, F., et al., (2015). Improving cold storage and processing traits in potato through targeted gene knockout. *Plant Biotechnology Journal*, *14*(1), 169–176. https://doi.org/10.1111/pbi.12370 (Accessed on 18 December 2019).
9. Coleman, K., (2017). *Genome Editing Could Stimulate Crop Yields*. https://www.pewtrusts.org/en/research-and-analysis/

articles/2017/12/04/genome-editing-could-stimulate-crop-yields (Accessed on 18 December 2019).

10. Curtin, S. J., Zhang, F., Sander, J. D., Haun, W. J., Starker, C., Baltes, N. J., e al., (2011). Targeted mutagenesis of duplicated genes in soybean with zinc-finger nucleases. *Plant Physiology*, *156*(2), 466–473. doi: 10.1104/pp.111.172981.
11. Demorest, Z. L., Coffman, A., Baltes, N. J., Stoddard, T. J., Clasen, B. M., Luo, S., Retterath, A., Yabandith, A., Gamo, M. E., Bissen, J., Mathis, L., Voytas, D. F., & Zhang, F., (2016). Direct stacking of sequence-specific nuclease-induced mutations to produce high oleic and low linolenic soybean oil. Creative commons attribution 4.0 international license. *BMC Plant Biology*, *16*(225). https://doi.org/10.1186/s12870-016-0906-1 (Accessed on 18 December 2019).
12. Fauser, F., Schiml, S., & Puchta, H., (2014). Both CRISPR/Cas-based nucleases and nickases can be used efficiently for genome engineering in *Arabidopsis thaliana*. *The Plant Journal*, *79*(2), 348–359. https://doi.org/10.1111/tpj.12554 (Accessed on 18 December 2019).
13. Haun, W., Coffman, A., Clasen, B. M., Demorest, Z. L., Lowy, A., Ray, E., Retterath, A., Stoddard, T., Juillerat, A., Cedrone, F., Mathis, L., Voytas, D. F., & Zhang, F., (2014). Improved soybean oil quality by targeted mutagenesis of the fatty acid desaturase 2 gene family. *Plant Biotechnology Journal*, *12*(7), 934–940. https://doi.org/10.1111/pbi.12201 (Accessed on 18 December 2019).
14. Li, M., Li, X., Zhou, Z., Wu, P., Fang, M., Pan, X., Lin, Q., Luo, W., Wu, G., & Li, H., (2016). Reassessment of the four yield-related genes *Gn1a*, *DEP1*, *GS3*, and *IPA1* in rice using a CRISPR/Cas9 system. Creative commons attribution license. *Frontiers in Plant Science, 7*, 377. doi: 10.3389/fpls.2016.00377.
15. Li, T., Liu, B., Spalding, M. H., Weeks, D. P., & Yang, B., (2012). High-efficiency TALEN-based gene editing produces disease-resistant rice. *Nature Biotechnology*, *30*, 390–392. doi:10.1038/nbt.2199.
16. Limera, C., Sabbadini, S., Sweet, J., & Mezzetti, B., (2017). New biotechnological tools for the genetic improvement of major woody fruit species. Creative commons attribution 4.0 international license. *Frontiers in Plant Science, 8*, 1418. doi: 10.3389/fpls.2017.01418.
17. Ma, L., Zhu, F., Li, Z., Zhang, J., Li, X., Dong, J., & Wang, T., (2015). TALEN-based mutagenesis of lipoxygenase LOX3 enhances the storage tolerance of rice (*Oryza sativa*) seeds. *PLoS One*, *10*(12),

e0143877. doi: 10.1371/journal.pone.0143877.

18. Ma, X., Zhang, Q., Zhu, Q., Liu, W., Chen, Y., Qiu, R., et al., (2015). A robust CRISPR/Cas9 system for convenient, high-efficiency multiplex genome editing in monocot and dicot plants. *Molecular Plant*, *8*(8), 1274–1284. doi: 10.1016/j.molp.2015.04.007.
19. Malzahn, A., Lowder, L., & Qi, Y., (2017). Plant genome editing with TALEN and CRISPR. Creative commons attribution international license. *Cell and Bioscience*, *7*, 21. doi: 10.1186/s13578-017-0148-4.
20. Mao, Y., Botella, J. R., Liu, Y., & Zhu, J. K., (2019). Gene editing in plants: Progress and challenges. Creative commons CC BY license. *National Science Review*, *6*(3), 421–437. https://doi.org/10.1093/nsr/nwz005 (Accessed on 18 December 2019).
21. Marius Walter. *Creative Commons Attribution-Share Alike 4.0 International License*. https://commons.wikimedia.org/wiki/File:DNA_Repair_after_CRISPR-Cas9_cut.svg; https://commons.wikimedia.org/wiki/File:GRNA-Cas9.png (Accessed on 18 December 2019).
22. Merck, (2019). *Mission™ Ribonucleoprotein (RNP) Protocols*. https://www.sigmaaldrich.com/technical-documents/articles/biology/ribonucleoprotein-based-geneome-editing.html (Accessed on 18 December 2019).
23. Metje-Sprink, J., Menz, J., Modrzejewski, D., & Sprink, T., (2019). DNA-free genome editing: Past, present and future. *Frontiers in Plant Science*, *9*, 1957. doi: 10.3389/fpls.2018.01957.
24. Miao, J., Guo, D., Zhang, J., Huang, Q., Qin, G., Zhang, X., Wan, J., Gu, H., & Qu, L. J., (2013). Targeted mutagenesis in rice using CRISPR-Cas system. *Cell Research*, *10*, 1233–1236. doi: 10.1038/cr.2013.123.
25. Ogletreerd. *Creative Commons Attribution-Share Alike 4.0 International License*. https://commons.wikimedia.org/wiki/File:Overview_of_TALENs.png (Accessed on 18 December 2019).
26. Osakabe, K., Osakabe, Y., & Toki, S., (2010). Site-directed mutagenesis in *Arabidopsis* using custom-designed zinc finger nucleases. *Proceedings of the National Academy of Sciences of the United States of America*, *107*(26), 12034–12039. doi: 10.1073/pnas.1000234107.
27. Osakabe, Y., & Osakabe, K., (2014). Genome editing with engineered nucleases in plants. *Plant and Cell Physiology*, *56*(3), 389–400. https://

doi.org/10.1093/pcp/pcu170 (Accessed on 18 December 2019).

28. Papaioannou, I., Simons, J. P., & Owen, J. S., (2012). Targeted *in situ* gene correction of dysfunctional APOE alleles to produce atheroprotective plasma ApoE3 protein. Creative commons attribution 3.0 unported license. *Cardiology Research and Practice 1*, 148796. http://dx.doi.org/10.1155/2012/148796 (Accessed on 18 December 2019).
29. Pater, S., Pinas, J. E., Hooykaas, P. J. J., & Zaal, B. J., (2013). ZFN-mediated gene targeting of the Arabidopsis *protoporphyrinogen oxidase* gene through *Agrobacterium*-mediated floral dip transformation. *Plant Biotechnology Journal*, *11*(4), 510–515. doi: 10.1111/pbi.12040.
30. Petolino, J. F., (2015). Genome editing in plants via designed zinc finger nucleases. Creative commons attribution license. *In Vitro Cellular and Developmental Biology-Plant*, *51*(1), 1–8. https://doi.org/10.1007/s11627-015-9663-3 (Accessed on 18 December 2019).
31. Petolino, J. F., Worden, A., Curlee, K., Connell, J., Strange, M. T. L., Larsen, C., & Russell, S., (2010). Zinc finger nuclease-mediated transgene deletion. *Plant Molecular Biology*, *73*(6), 617–628. https://doi.org/10.1007/s11103-010-9641-4 (Accessed on 18 December 2019).
32. Shan, Q., Zhang, Y., Chen, K., & Gao, C., (2015). Creation of fragrant rice by targeted knockout of the OsBADH2 gene using Talen technology. *Plant Biotechnology Journal*, *13*(6), 791–800. https://doi.org/10.1111/pbi.12312 (Accessed on 18 December 2019).
33. Shi, J., Gao, H., Wang, H., Lafitte, H. R., Archibald, R. L., Yang, M., Hakimi, S. M., Mo, H., & Habben, J. E., (2016). ARGOS8 variants generated by CRISPR-Cas9 improve maize grain yield under field drought stress conditions. Creative commons attribution 4.0 international license. *Plant Biotechnology Journal*, *15*(2), 207–216. https://doi.org/10.1111/pbi.12603 (Accessed on 18 December 2019).
34. Shukla, V. K., Doyon, Y., Miller, J. C., DeKelver, R. C., Moehle, E. A., Worden, S. E., et al., (2009). Precise genome modification in the crop species *Zea mays* using zinc-finger nucleases. *Nature*, *459*, 437–441. doi: 10.1038/nature07992.
35. Sun, Y., Zhang, X., Wu, C., He, Y., Ma, Y., Hou, H., Guo, X., Du, W., Zhao, Y., & Xia, L., (2016). Engineering herbicide-resistant rice plants through CRISPR/Cas9-mediated homologous recombination of acetolactate synthase. *Molecular Plant*, *9*(4), 628–631. https://doi.org/10.1016/j.molp.2016.01.001 (Accessed on 18 December 2019).

36. Tang, X., Zheng, X., Qi, Y., Zhang, D., Cheng, Y., Tang, A., Voytas, D. F., & Zhang, Y., (2016). A single transcript CRISPR-Cas9 system for efficient genome editing in plants. *Molecular Plant*, *9*(7), 1088–1091. https://doi.org/10.1016/j.molp.2016.05.001 (Accessed on 18 December 2019).
37. Townsend, J. A., Wright, D. A., Winfrey, R. J., Fu, F., Maeder, M. L., Joung, J. K., & Voytas, D. F., (2009). High frequency modification of plant genes using engineered zinc finger nucleases. *Nature*, *459*(7245), 442–445. doi: 10.1038/nature07845.
38. Woo, J. W., Kim, J., Kwon, S. I., Corvalán, C., Woo, S. W., Kim, H., Kim, S. G., Kim, S. T., Choe, S., & Kim, J. S., (2015). DNA-free genome editing in plants with preassembled CRISPR-Cas9 ribonucleoproteins. *Nature Biotechnology*, *33*, 1162–1164. doi: 10.1038/nbt.3389.
39. Xie, K., Minkenberg, B., & Yang, Y., (2015). Boosting CRISPR/Cas9 multiplex editing capability with the endogenous tRNA-processing system. *Proceedings of the National Academy of Sciences of the United States of America*, *112*(11), 3570–3575. https://doi.org/10.1073/pnas.1420294112 (Accessed on 18 December 2019).
40. Zhang, F., Maeder, M. L., Unger-Wallace, E., Hoshaw, J. P., Reyon, D., Christian, M., Li, X., Pierick, C. J., Dobbs, D., Peterson, T., Juong, J. K., & Voytas, D. F., (2010). High frequency targeted mutagenesis in *Arabidopsis thaliana* using zinc finger nucleases. *Proceedings of the National Academy of Sciences of the United States of America*, *107*(26), 12028–12033. https://doi.org/10.1073/pnas.0914991107 (Accessed on 18 December 2019).
41. Zhang, Y., Liang, Z., Zong, Y., Wang, Y., Liu, J., Chen, K., Qiu, J. L., & Gao, C., (2016). Efficient and transgene-free genome editing in wheat through transient expression of CRISPR/Cas9 DNA or RNA. *Nature Communications*, *7*, 12617. doi: 10.1038/ncomms12617.
42. Zhang, Y., Liang, Z., Zong, Y., Wang, Y., Liu, J., Chen, K., Qiu, J. L., & Gao, C., (2017). Efficient and transgene-free genome editing in wheat through transient expression of CRISPR/CAS9 DNA or RNA. Creative commons attribution 4.0 international license. *Nature Communications, 7*, 12617. doi: 10.1038/ncomms12617.
43. Zhang, Y., Massel, K., Godwin, I. D., & Gao, C., (2018). Applications and potential of genome editing in crop improvement. Creative commons attribution 4.0 international license. *Genome Biology*, *19*, 210. doi: 10.1186/s13059-018-1586-y.

INDEX

E

F

G

H

Q

R

S